Advanced Simulation Methods for ESD Protection Development

Full catalogue information on all books, journals and electronic products can be found on the Elsevier Science homepage at: http://www.elsevier.com

ELSEVIER PUBLICATIONS OF RELATED INTEREST

JOURNALS:

Advanced Engineering Informatics
Automatica
Computers and Electrical Engineering
Computers in Industry
Control Engineering Practice
Engineering Applications of AI
Journal of Electrostatics
Journal of Manufacturing Systems
Journal of Materials Processing Technology
Journal of the Franklin Institute
Measurement
Mechatronics
Microelectronics Reliability
Microelectronics Journal
Robotics and Autonomous Systems
Robotics and Computer-Integrated Manufacturing
Sensors and Actuators
Signal Processing

To contact the Publisher

Elsevier Science welcomes enquiries concerning publishing proposals: books, journal special issues, conference proceedings, etc. All formats and media can be considered. Should you have a publishing proposal you wish to discuss, please contact, without obligation, the publisher responsible for Elsevier's Control, Electronic and Optical Engineering programme:

Mr Christopher Greenwell
Publishing Editor
Elsevier Science Ltd
The Boulevard, Langford Lane   Phone: +44 1865 843230
Kidlington, Oxford   Fax: +44 1865 843920
OX5 1GB, UK   E.mail: c.greenwell@elsevier.com

General enquiries, including placing orders, should be directed to Elsevier's Regional Sales Offices – please access the Elsevier homepage for full contact details (homepage details at the top of this page).

Book Butler logo to search for more Elsevier books, visit the Books Butler at http://www.elsevier.com/homepage/booksbutler/

Advanced Simulation Methods for ESD Protection Development

**Kai Esmark, Harald Gossner and
Wolfgang Stadler**
Infineon Technologies, Munich, Germany

2003

ELSEVIER

**Amsterdam - Boston - Heidelberg - London - New York - Oxford
Paris - San Diego - San Francisco - Singapore - Sydney - Tokyo**

ELSEVIER Ltd
The Boulevard, Langford Lane
Kidlington, Oxford OX5 1GB, UK

First edition 2003

Library of Congress Cataloging in Publication Data
A catalog record from the Library of Congress has been applied for.

British Library Cataloguing in Publication Data
A catalogue record from the British Library has been applied for.

ISBN: 0-08-044147-5

♾ The paper used in this publication meets the requirements of ANSI/NISO Z39.48-1992 (Permanence of Paper).
Printed in the Netherlands.

TO OUR FAMILIES

Anne and Merle ESMARK

Ursula, Thomas and Antonia GOSSNER

Eva–Maria, Andreas, Michael and Katharina STADLER

Preface

The goal of this book is to introduce the simulation methods necessary to describe the behaviour of semiconductor devices during an electrostatic discharge (ESD). The challenge of this task is the correct description of semiconductor devices under very high current density and high temperature transients. As it stands, the book can be no more than a snapshot and a summary of the research in this field during the past few years. The authors hope that the book will provide the basis for further development of simulation methods at this current frontier of device physics.

The book is intended both for engineers responsible for the ESD protection development of circuits in the industrial environment, who can closely follow the given examples to solve their specific problems, and for readers who are more generally interested in basic methodology, such as students attending advanced courses in microelectronics and academics working in the field of transport simulation.

The book presents a complete simulation flow which can be applied to ESD protection development. First, the reader is introduced to the phenomenon of ESD, including the problems and questions posed by ESD protection development. The common concepts of ESD protection are presented and the parameters which influence ESD performance are traced. After an overview of the simulation flow and simulation tools, the construction of devices under investigation by process simulation is described. These devices are "operated" by device simulation and the electrical data are acquired under conditions of very high current densities and temperatures in the device. These electrical data are correlated with the ESD parameters and the ESD performance of the device. The book shows how equivalent circuits and the corresponding parameters for the devices can be extracted from the electrical data acquired during device simulation. These "compact models" are fed into a circuit simulation, which allows the ESD simulation of a greater number of devices, such as are typically found in the I/O (input/output) cells of an integrated circuit (IC). As the ESD discharge into an IC involves a large part of the IC like the metal bus system serving as "lightning conductor", the requirements and approaches of a full IC simulation are analysed. In each chapter a summary of the main technical results is given.

This is an appropriate place to acknowledge all the scientific work described in the book, which has been performed over recent years in industry, universities and research institutions world-wide. We make special acknowledgement of the ESD and latch-up protection development team at Infineon Technologies, Munich, Germany, where many of the presented results were recorded and verified in numerous technologies over many years. Our successful collaborations with the institutes of Prof. Wolfgang Fichtner at the Swiss Federal Institute of Technology, Zurich and Prof. Erich Gornik at the Vienna University of Technology, the team of Dr Horst Gieser at the Fraunhofer Institute, Munich, Prof. Winfried Soppa of the University of Applied Science Osnabruck, and Prof. Doris Schmitt-Landsiedel of the Technical University Munich, significantly contributed to the new findings discussed in the book. Some of the results have been obtained under the auspices of the European and German research projects PARASITICS and ASDESE, funded by the German Federal Ministry of Education and Research (BMBF). We are grateful for the consistent and dedicated support from the failure analysis group at Infineon Technologies, from where the failure pictures presented in the book were taken. Thanks also belongs to the members of the ISE AG (Zurich, Switzerland) for their quick and continuous support in case of questions regarding the simulation tools. Without the detailed and thoughtful reviews of Dr Tilo Brodbeck, Heinz Endriss, Xaver Guggenmos and Dr Martin Streibl (Infineon Technologies), Stefan Drüen (TU Munich), Prof. Winfried Soppa (University of Applied Science Osnabruck), Heinrich Wolf (Fraunhofer Institute, Munich), Martin Litzenberger and Dr Dionyz Pogany (TU Vienna), all experts in their fields, and the extensive technical support of Jens Weissenburger (Infineon Technologies), the book would not have achieved its present standard. We would also like to thank the editor, Dr Martin Ruck, for good and supportive collaboration, and Prof. Ninoslav Stojadinovic for support during the initiation of this project. We are grateful for the positive and creative atmosphere at Corporate Development of Infineon Technologies, which allowed us to realize this project. However, we would have failed to finish this book without the patience and the motivation of our families. Therefore our biggest debt is to them.

Kai Esmark
Harald Gossner
Wolfgang Stadler

Munich, April 2003

Contents

Chapter 1

ESD basics

1.1 Introduction

The effect of electrostatic discharge (ESD) is an everyday experience. Even before you enter the office you can be affected by ESD when you step out of your car and touch the car body. It is tangible, sometimes even hurtfully tangible, which suggests there is power behind it. In fact, the power calculated according the physical laws can amount to several kilowatts. The reason why you can nevertheless reach your desk in the morning without serious harm is that the duration of the discharge, in other words the length of the discharge pulse, is quite short, typically between a few to some hundred nanoseconds. Therefore the dissipated "destructive" energy is low compared to when, for example, somebody steps on your toes.

However, considering the possible damage it might be misleading to speak about low pulse energy, because the size of the system hit by the discharge is equally important. Even if there is no noticeable damage to a human being or a car body, the same pulse can seriously affect the car if it hits the contact of the motor control integrated circuit (IC). The density of the dissipated energy is the most important parameter, and this density can reach values of 10^3 J/cm^3 for modern integrated circuits directly hit by an ESD event with an energy density comparable to that when lightning strikes a big tree. Considering the electronic components during an ESD event, we find semiconductor devices operating under extraordinary conditions with current densities up to 100 mA per square micrometer and temperatures close to the melting point of the semiconductor material. Even this stress can be survived because of the short pulse length, if certain measures are taken to prepare the critical part of the circuitry for the appropriate conditions. This is the essential motivation of ESD protection development for ICs.

One can now imagine that a simulation under such conditions requires consideration of the physical effects, and their careful calibration beyond the well-known models of integrated circuits during normal operation. In the course of this book we will trace the parameters, physical models and the conditioning of the simulation tools which are necessary to perform a successful and reliable simulation under these high current density and high temperature conditions. Besides ESD, any other transport simulation considering high temperature transients in a semiconductor device will encounter similar challenges and can benefit from the described electro-thermal coupling in the transport modelling and the extraction of the various temperature dependent model parameters at high temperatures.

Returning to the ESD phenomenon, there are several excellent reference books which provide information on electrostatics (Greason, 1992), ESD models (Bhar, 1983), ESD protection methods (Diaz, 1995; Matisoff, 1986; Corp, 1990), design methodology (Dabral, 1998), ESD device physics (Amerasekera, 2002; Antinone, 1986), and external ESD protection measures (Baumgärtner, 1997). A recent overview of these topics is also given in Wang (2002).

The task of ESD protection development for ICs is to determine the technological and design measures which will allow the IC to withstand ESD pulses up to a certain energy or maximum voltage. Beyond this level it is required that external (usually costly) measures, such as appropriate grounding of operators and machines or discrete protection elements on the board or within the system, limit the possible charging or discharging.

The first issue for ESD protection development is to characterize the robustness of an integrated circuit. This needs an abstraction of the real world conditions where pulse waveforms of any kind may appear. A good comparison of protection techniques is only possible by using well-defined discharge pulses, since parameters like rise time, peak current and total energy all play a role. The most common standards are discussed in Chapter 1.2. Of course, the selected test method should be representative for fails appearing under real world conditions. In other words, it must address typical ESD failure mechanisms and locations (see Chapter 1.3). The goal for ESD protection development is to achieve the intended ESD robustness level with a minimum of extra area and performance degradation under normal operation. There are many ways to optimize the robustness, for example the use of specific devices at critical locations in the circuit, a change of the layout of endangered devices, a change in process technology, or all of these. Analysis methods which help to select the best protection approach for the considered circuit and process technology are described in Section 1.4.

1.2   The ESD event – models and testing

1.2.1   Overview

Electrostatic discharge is commonly defined as a transient current flow compensating charge imbalance between bodies. The natural phenomenon of lightning is probably the most impressive and well-known example of ESD. In lightning, the air between the charged clouds and the earth (ground) breaks down if the breakdown field strength is exceeded. A conductive path is formed by a plasma of ionized gas molecules. The resulting current flow can be seen as a flash. The same phenomenon, on much smaller scale with respect to the initial electrostatic voltage and the resulting breakdown lengths, can occur during the handling of integrated circuits and microelectronic equipment. If no appropriate protection measures exist, the result of both ESD phenomena will be destructive.

During the life of an IC, the risk of damage by an electrostatic discharge or electrical overstress (EOS) by surges is a constant threat. Starting in the wafer fab during processing, charge can be introduced to the wafer, for example by etching steps with acid. From the many case studies in the literature it is known that the handling of silicon dies during assembly, as well as the handling of packaged ICs by non-safely grounded operators or automatic equipment, can be severe yield detractors (Dangelmayer, 1990; Gärtner, 1999; Yan, 2001). Even in a system in which the IC is mounted on a printed circuit board (PCB) and superficially well shielded by an anti-static case, ICs might be exposed to surges causing EOS, particularly if the system is used in a rough environment (e.g. system-level ESD in automotive products) or if the IC has a critical interface to the external world (e.g. lightning surges in wireline communication ICs).

Although the source of the electrostatic discharge may be totally different, the mechanisms which produce hazard for the ICs are very similar, regardless of whether the victim of the ESD shock is a single die, a packaged IC or an entire system. In what follows, we will focus mainly on ESD events and protection of single packaged ICs.

There are two different categories of electrostatic discharge. In the first category, the device (or die, board, electronic system ...) is touched by a charged human being or machine which discharges to ground via the electronic device. The stress can be characterized by a two terminal model: the charge is forced from one terminal (pin) of the device to at least one other pin which is connected to ground. The resulting current is in first order determined simply by the charge stored on the discharging capacitor and the total resistance in the discharge path. In the second ESD category, the capacitance of an IC itself charges up or discharges in order to balance its charge with the environment. In the worst case, the discharge occurs through one single pin. In contrast to the stress models in the first category, the discharge current here depends on the IC itself, i.e., on the package, which largely determines the capacitance, and the environment, which influences the formation of the discharge arc.

For more than 20 years, models to reproduce real-world ESD events have been discussed and refined by several standardization organizations. The goal of such test models is obvious. If the model describes the stress which could occur in reality, the robustness of a product according to that test model is a reliable measure for the risk of damage to the product in the field. Currently, there are three basic ESD models covered by standards: Human Body Model (HBM), Machine Model (MM), and variations of the Charged Device Model (CDM). Simple representations of these discharges and the corresponding basic lumped element circuits are depicted in Figure 1.1. All these models can be described in first order by a simple RLC network (see, for example, Gieser (2002), and references therein). The second order differential equation of the basic RLC networks of Figure 1.1 with series resistance R_{ESD}, which is a sum of the resistors within the ESD model and the resistor of the load under ESD conditions, R_{L}, the capacitor C_{ESD}, initially

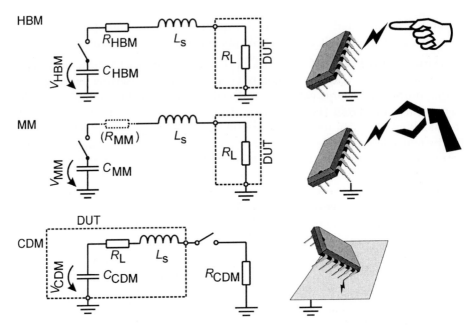

Figure 1.1: Examples of discharge according to HBM, MM, and CDM and its representation by simple RLC circuits. The typical parameters for the various discharge types are summarized in Table 1.1.

charged up to a voltage V_{ESD}, and the series inductance L_{s} in the discharge path

$$L_{\mathrm{s}}\frac{\mathrm{d}^2 i}{\mathrm{d}t} + R_{\mathrm{ESD}}\frac{\mathrm{d}i}{\mathrm{d}t} + \frac{1}{C_{\mathrm{ESD}}}i = 0 \qquad (1.1)$$

can be solved analytically:

$$I_{\mathrm{ESD}}(t) \;=\; V_{\mathrm{ESD}}C_{\mathrm{ESD}}\frac{\omega_0^2}{\sqrt{\alpha^2-\omega_0^2}}\,\mathrm{e}^{-\alpha t}\sinh\left(\sqrt{\alpha^2-\omega_0^2}\,t\right) \quad \text{for } \alpha > \omega_0 \quad (1.2)$$

$$I_{\mathrm{ESD}}(t) \;=\; V_{\mathrm{ESD}}C_{\mathrm{ESD}}\frac{\omega_0^2}{\sqrt{\omega_0^2-\alpha^2}}\,\mathrm{e}^{-\alpha t}\sin\left(\sqrt{\omega_0^2-\alpha^2}\,t\right) \quad \text{for } \alpha < \omega_0 \quad (1.3)$$

with the damping coefficient $\alpha = R_{\mathrm{ESD}}/2L_{\mathrm{s}}$ and the oscillation frequency $\omega_0 = 1/\sqrt{L_{\mathrm{s}}C_{\mathrm{ESD}}}$. Examples for typical parameters of the three basic models are compared in Table 1.1. The equivalent pulse shapes of the discharges are plotted in Figure 1.2. Besides these ESD models, square pulse test methods with various pulse length are widely used for analysis of protection capabilities of structures and circuits, and, therefore, are also valuable tools for simulation studies (see Section 1.4.1).

There are several publications which discuss the models, with all their advantages and deficiencies, in great detail (for an excellent summary see Gieser, 2002). Therefore, in the following sections, we will focus on the aspects of the models which are relevant for simulation purposes.

Table 1.1: Examples of typical parameters for the RLC circuits in Figure 1.1 and the resulting typical peak currents, the 10 % to 90 % rise times, the parameters α and ω_0 defined in Equations 1.3 and 1.2, and the FWHM of the first peak of the discharge pulse.

Parameter	HBM	MM	CDM
V_{ESD} (example)	4000 V	200 V	500 V
$R_{\text{HBM}}/R_{\text{MM}}/R_{\text{CDM}}$	1.5 kΩ	5 Ω	10 Ω
C_{ESD}	100 pF	200 pF	10 pF
L_{s}	5000 nH	750 nH	10 nH
R_{L}		10 Ω	
I_{ESD}	2.6 A	2.8 A	10.4 A
t_{rise} (10 %/90 %)	≈ 7 ns	≈ 11 ns	≈ 0.3 ns
α	1.5×10^8 s^{-1}	0.1×10^8 s^{-1}	10×10^8 s^{-1}
ω_0	0.5×10^8 s^{-1}	0.8×10^8 s^{-1}	30×10^8 s^{-1}
FWHM (1st peak)	≈ 120 ns	≈ 26 ns	≈ 0.7 ns

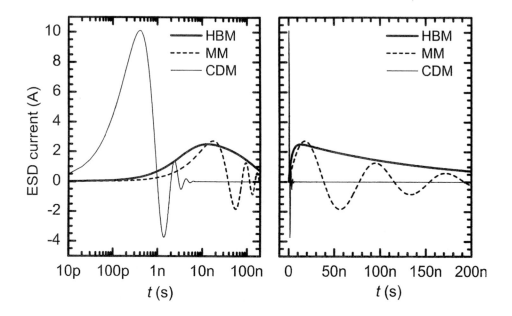

Figure 1.2: Typical discharge pulse shapes for the RLC networks of the HBM, MM, and CDM discharge in logarithmic (left) and linear time scale (right). The adopted parameters are listed in Table 1.1.

1.2.2   Human Body Model

Without a doubt, the Human Body Model is today the most commonly used discharge model in the microelectronic industry. The intention of the model is to reproduce a discharge of a charged human being to a device with at least one pin grounded. Although the risk of an IC getting touched by a charged human being has decreased significantly due to automatic handling, the HBM is relevant because it is – in comparison to the other models – well defined by international standards. Based on the military standard MIL–STD–883C method 3015.7 from 1989 (MIL–STD–HBM, 1989), several international standard committees have extended and refined the standard and the corresponding test instruction, e.g. JEDEC–HBM (2000), ESDA–HBM (2001), IEC–HBM (2002). It should be noted that similar discharge mechanisms, i.e. slow discharge, also exist in fully automated production lines.

In first order, the human being is reduced to a 100 pF capacitor C_{HBM} which is charged up to V_{HBM}, and a serial resistance of $R_{\mathrm{HBM}} = 1.5$ kΩ. Typical charging voltages range from one to several kilovolts. Usually, at these stress voltages there is either a breakdown of pn junctions or a dielectric breakdown in the IC. The breakdown of pn junctions is in general also the protection mechanism, offering a low ohmic discharge path on the IC. In this case the HBM discharge model forces a current I_{HBM} through the device. The threshold current (or withstand current) is the essential parameter describing the robustness. Consequently, throughout the book stress levels and failure modes will be discussed in terms of ESD current.

The discharge of that simple RC network results in an exponential decay of the current

$$I_{\mathrm{HBM}}(t) = \frac{V_{\mathrm{HBM}}}{R_{\mathrm{HBM}}} \exp\left(\frac{-t}{R_{\mathrm{HBM}} C_{\mathrm{HBM}}}\right) \tag{1.4}$$

if R_{L} is small compared to R_{HBM}. The parasitic series inductances L_{s} in the discharge path prevent an ideal decay of an instantly flowing current and determine a rise time for the pulse.

For the HBM, typical values for L_{s} of ESD testers are in the range of 5– 10 µH (Verhaege, 1993). Therefore, $\alpha = 1.5 \times 10^8$ s$^{-1} > \omega_0 = 0.5 \times 10^8$ s^{-1}, and differentiation of the simplified Equation 1.2 leads to a 0 %/100 % rise time of

$$t_{\mathrm{rise}} = \frac{1}{2\sqrt{\alpha^2 - \omega_0^2}} \ln\left(\frac{\alpha - \sqrt{\alpha^2 - \omega_0^2}}{\alpha + \sqrt{\alpha^2 - \omega_0^2}}\right) \tag{1.5}$$

This can be approximated if α is comparable to ω_0 by

$$t_{\mathrm{rise}} \approx \frac{2L_{\mathrm{s}}}{R_{\mathrm{HBM}}} \tag{1.6}$$

However, the RLC circuit is much too simplistic to represent a real HBM tester and cannot correctly reproduce discharge waveforms from real testers. While the discharge in a short can be described sufficiently by the simple RLC model, for loads ($R_{\mathrm{L}} > 200$ Ω) this simple model fails. Roozendaal (1990) and Verhaege (1993) developed a more realistic equivalent sub-circuit for HBM testers by adding

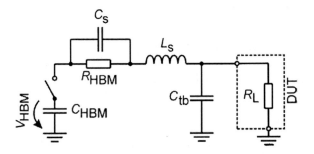

Figure 1.3: Equivalent HBM sub-circuit including the HBM resistor, R_{HBM}, the HBM capacitance, C_{HBM}, the series inductance in the discharge path, L_{s}, and the lumped elements for the stray capacitance across the HBM resistor, C_{s}, and the test board capacitance, C_{tb}. The device under test is represented by a (load) resistor, R_{L} (Roozendaal, 1990; Verhaege, 1993).

tester parasitics as lumped elements, namely the stray capacitance of the HBM resistor, C_{s}, and the test board capacitance, C_{tb} (Figure 1.3). Additionally, for an exact analysis of the current waveform, the resistance of the device under test, R_{L} (load resistance), must be considered. Verhaege (1993) showed that for the resulting sub-circuit of Figure 1.3 there exists an analytical solution which fits HBM waveforms obtained experimentally.

For the determination of the parasitics of a specific HBM tester, it is important that the calibration measurements be performed not only with a short circuit as required in the MIL standard (MIL–STD–HBM, 1989), but also with a 500 Ω load as defined in the newer standards (JEDEC–HBM, 2000; ESDA–HBM, 2001; IEC–HBM, 2002). Obviously, there is a rather complex interaction between the lumped element model of the HBM tester and the device under test (DUT).

- For a short circuit as calibration element of the HBM tester, the board capacitance C_{tb} does not play any significant role. The stray capacitance across R_{HBM} increases the peak current I_{HBM} and reduces t_{rise}.

- With increasing resistive loads, the test board capacitance C_{tb} becomes more important. For a simple resistive load, C_{tb} decreases the peak current and therefore "weakens" the ESD test and increases the rise time.

- According to the discussion in Roozendaal (1990), the test board capacitance C_{tb} discharges at any snap-back point in the IV characteristic. For ESD protection devices based on a non-destructive snap-back, such as silicon controlled rectifiers (SCRs) and (parasitic) bipolars, this "extra" current due to the discharge of the C_{tb} can be hazardous. In real testers, this effect may be diminished because the test board capacitance is distributed, instead of being a single lumped element (Russ, 1994).

The experimentally determined HBM pulse shapes of two commercially available HBM testers are shown in Figure 1.4 for a short-circuit load and a 500 Ω resistive load. A fit with a simple network containing R_{HBM}, C_{HBM}, and an

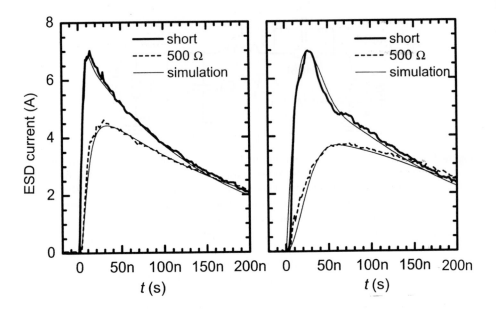

Figure 1.4: Typical waveforms for a standard HBM tester (left) and a wafer-level HBM
         tester (right) for a discharge with $V_{HBM} = 1$ kV into a short and into a 500 Ω
         resistor.

inductance L_s does not reproduce the pulse shape sufficiently well. From the
short-circuit measurements, C_s is determined. Knowing R_{HBM}, C_{HBM}, L_s, and
C_s, the test board capacitance can be obtained from the calibration measurements
using a 500 Ω resistor load. The extracted lumped elements are summarized in
Table 1.2. Tester 1 is compliant with current standards. Tester 2, an HBM wafer
level tester, meets the requirements for zero load, but fails the requirements of
the IEC, JEDEC, and ESDA standard concerning the 500 Ω load. Obviously, as
depicted in Figure 1.4, the simulations fit the pulse shapes of the HBM standard
tester quite well using the parameters in Table 1.2. For the wafer level tester
larger, but for most applications not critical, deviations between simulation and
experimentally obtained data must be accepted. This deviation probably indicates
that the model of Figure 1.4 is still not sufficient for the wafer level tester with
its specific configuration, i.e. rather long cable connections from the test board to
the probe card.

It should be mentioned here that, although the recent standards describe the
HBM equivalent circuit quite well, there are several open issues concerning the
correlation of HBM testers. The main problem seems to be the rise time of the
HBM pulse which is, in the most recent standards, defined to be between 2–10 ns
from 10 % to 90 % of the maximum HBM current I_{HBM}. Several publications have
addressed the effect of the rise time on the ESD robustness of single protection
devices as well as on complete integrated circuits (Mußhoff, 1996; Barth, 2000).

In addition to the Human Body Model on a component level, there is a standard

Table 1.2: HBM tester parameters according to the equivalent circuit in Figure 1.3 for the HBM discharges shown in Figure 1.4. Data taken with kind permission from T. Brodbeck and S. Drüen.

tester	R_{HBM}	L_{s}	C_{s}	C_{tb}
HBM standard	1500 Ω	8 μH	1.5 pF	28.6 pF
HBM wafer-level	1550 Ω	20 μH	5.5 pF	80 pF

defining a Human Body Model for system level applications (EN–61000, 1996). This standard defines a discharge of a 150 pF capacitor into a 330 Ω resistor. In contrast to the well-defined HBM standard for component level testing, the EN 61000–4–2 specifies only certain points of the discharge pulse shape of the ESD simulator. There are several crucial factors, such as the connection of the ground to the test set-up, which have a severe influence on the discharge wave form; these, however, are not defined in the standard.

1.2.3 Machine Model

Like HBM, the machine model represents the discharge of a capacitor into a component which has at least one pin grounded. The main difference from the HBM is that the capacitor is discharged into the device via a very low impedance. Additionally, the discharging capacitance is increased to 200 pF, twice the value of the HBM. These changes result in a much more severe stress at the same charging voltages for the MM compared to the HBM. Consequently, the failure threshold voltage is 5–20 times smaller than the HBM (see also Table 1.1). However, in general the same failure mechanism is addressed. Therefore, the MM is often called "worst–case HBM".

The main problem with the MM is the demand for a $R_{\mathrm{MM}} = 0$ Ω, which of course cannot be achieved in real testers. In reality there is a significant impedance during the discharge, for example due to the test board, the relay matrix, the fixture, and the connections in between. Nevertheless, the resulting parasitic resistance is in many cases still smaller than the resistance of the device under test. Therefore, the test device itself has significant influence on the discharge waveforms and can no longer be neglected. In addition, the parasitic series inductance L_{s} plays such a significant role that a comparison of the results from different testers can hardly ever be made. According to Roozendaal (1990), the damping coefficient α in Equation 1.2 and 1.3 must consider the resistance of the DUT.

$$\alpha = \frac{R_{\mathrm{MM}} + R_{\mathrm{L}}}{2 L_{\mathrm{s}}} \qquad (1.7)$$

where R_{MM} accounts for the total parasitic impedance of the test system. The shape of the discharge waveform depends on the ratio α/ω_0.

- For $\alpha > \omega_0$ a waveform similar to an HBM event is obtained (see Equation 1.2). The MM waveform shows a fast exponential increase, followed by

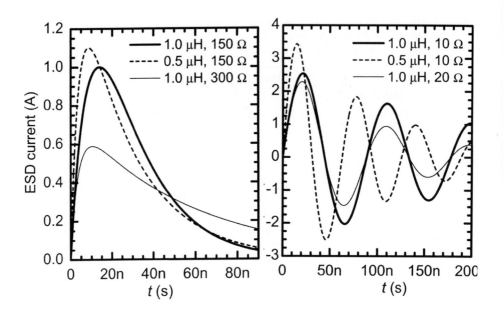

Figure 1.5: Current waveforms for MM discharges ($V_{\mathrm{MM}} = 200$ V) with varying parameters for R_{L} and L_{s}. Left: over-damped waveform similar to HBM ($\alpha > \omega_0$);
right: oscillating waveform ($\alpha < \omega_0$).

an exponential decay. The rise time and amplitude depend on the parameters
as discussed in Section 1.2.2.

- For $\alpha < \omega_0$ a MM discharge leads to the "typical" oscillating MM waveform
with a maximum peak current of $I_{\mathrm{MM}} = V_{\mathrm{MM}}\sqrt{C_{\mathrm{MM}}/L_{\mathrm{s}}}$, an oscillation
frequency of ω_0 and exponential damping with the damping factor α (see
Equation 1.3). Often the MM parameters in the equivalent circuit fulfil
$\omega_0 \gg \alpha$. In this case, Equation 1.3 reduces to

$$I_{\mathrm{MM}}(t) = V_{\mathrm{MM}}\sqrt{\frac{C_{\mathrm{MM}}}{L_{\mathrm{s}}}}\, \mathrm{e}^{-\alpha t} \sin\left(\frac{t}{\sqrt{L_{\mathrm{s}}C_{\mathrm{MM}}}}\right) \qquad (1.8)$$

The dependence of the MM discharge waveforms on L_{s} and R_{MM} is shown in Figure 1.5. Discharges with parasitic parameters where $\alpha > \omega_0$ result in over-damped
waveforms. Here R_{L} has a more severe influence on the discharge waveform than
L_{s}.

    In the case $\alpha < \omega_0$, oscillating waveforms are obtained. L_{s} then becomes the
decisive parameter which significantly influences the oscillation frequency, rise time
and peak current. A larger R_{L} has only a small effect on the amplitude of the first
peak, but it increases the damping coefficient α. The strong impact of L_{s} is also
a very critical issue for MM analysis at the wafer level (Yokoi, 2000) where the
relatively long cables between the test head and the probe card add a significant
parasitic inductance.

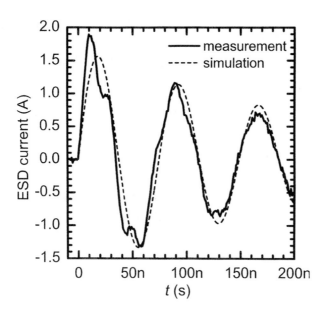

Figure 1.6: Current waveform for a 100 V MM discharges into a short. The parameters of the elements in the equivalent circuit of Figure 1.1 are $L_s = 700$ nH, $R_{MM} = 6$ Ω, $C_{MM} = 200$ pF. Data taken with kind permission from T. Brodbeck.

Owing to the strong dependence of the peak current and rise time on these parameters, their values have been specified in recent standards. MM waveforms defined by IEC (IEC–MM, 2002), JEDEC (JEDEC–MM, 1998) as well as ESDA (ESDA–MM, 1998) can be modelled by an equivalent RLC circuit using a parasitic resistance in the discharge path of $R_{MM} = 10$ Ω and a parasitic series inductance of $L_s = 750$ nH. The current standards also define a discharge in a 500 Ω load. For the MM, a discharge into a 500 Ω load is particularly important, because the parasitic impedance can be deduced from the waveform. For a discharge into a short, the variation of R_{MM} can be balanced by a change of L_s, leading to different peak currents for real devices with $R_L > 0$. Only the definition of a 500 Ω discharge waveform guarantees reproducible results. However, even for testers that comply with current standards, the issue of correlation between different testers and even between different DUTs measured by the same tester is much more severe for the MM than for the HBM. Together with the fact that MM addresses the same failure modes in the device as HBM (Amerasekera, 2002), testing with the HBM seems the preferable method for characterizing the ESD sensitivity of components.

For simulation purposes, the RLC model may be sufficient to reproduce the discharge waveform. An example of a discharge into a short and the corresponding calculated waveforms according to the RLC network is shown in Figure 1.6. Stray capacitance and test board capacitance can be neglected for this particular tester.

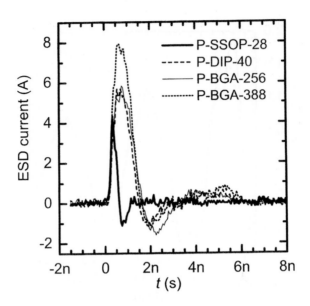

Figure 1.7: Current waveform for 500 V CDM discharges of "daisy chains" (empty packages with pins bonded to the lead frame) in different packages. The differences
in peak current as well as in pulse shape are obvious. (Data taken with kind
permission from T. Brodbeck.)

1.2.4 Charged Device Model

Finding an appropriate equivalent circuit for a discharge according to the Charged
Device Model which describes the real world CDM event is exceedingly difficult.
The reason for the difficulties is in the nature of the CDM.

The CDM phenomenon describes either the charging of a previously grounded
device, or the discharging of an initially charged device through a low-ohmic path.
In contrast to the HBM, the discharge waveform of a CDM event depends in
a complex manner on the IC and its environment. While for HBM and MM
discharges a well-defined capacitor of the tester defines the stored charge, in a
CDM event the amount of the transferred charge is determined by the device
itself. The capacitance of a device is a sum of several distributed capacitances,
such as the capacitance of the package, wiring, substrate, pad etc. to ground. Here
it should not be overlooked that also the position of the IC relative to the ground
plane is important. With the discharge waveform, the peak current amplitude,
and consequently the CDM withstand voltage, are changed. As an example, the
influence of the package on the discharge is shown in Figure 1.7.

There is still a debate about the right way to test integrated circuits with CDM
and how to standardize it. Current standards are available from JEDEC (JEDEC–
CDM, 2000) for the so-called field-induced CDM based on the ideas of Renninger
(1989) and the ANSI ESDA Standard (ESDA–CDM, 1999) for field-induced CDM

and direct-charging CDM. Additionally, there is a technical report for the ESDA standard (ESDA–SDM, 2000) for the Socketed Device Model (SDM) which is an attempt to reproduce CDM events by discharge of a socket device across relays (see, for example, Brodbeck 1998). A very detailed discussion of existing test methodologies can be found in Gieser (1999). On the one hand, the test should be able to reproduce the "real world discharge" in the field as closely as possible. On the other hand, the test must be reproducible and should be as easy to handle and cost-saving as possible. There is certainly still a significant lack of understanding of the physical mechanisms during a CDM event, particularly the influence of the discharge arc on the waveforms in the non-socketed methods need further detailed investigations. An additional challenge is the need for elaborate and costly test equipment to capture the very fast pulse.

Using a simple RLC network as illustrated in Figure 1.1, the discharge waveform for CDM is normally given by the oscillating solution, Equation 1.3, of the second order differential equation, Equation 1.1. Here, $C_{ESD} = C_{CDM}$ is the sum of all capacitances in the device (wiring, substrate, ...) and the package (pins) with respect to ground. C_{CDM} is often called background capacitance C_{back}. R_{CDM} accounts for the whole resistance in the discharge path, i.e. the resistance of the tester, the resistance of the arc and the total resistance of the discharge path in the DUT. This simple approach has been successfully used to reproduce experimentally determined CDM pulse shapes (Russ, 1996; Beebe, 1998). In Figure 1.8 an experimentally obtained current waveform together with the numerical fit using an RLC network is shown. For at least the first positive and the first negative peak, the discharge current simulated with the RLC network fits the experimental data well. In contrast, an SDM pulse cannot be described well by that simple model. Only a rough estimate can be obtained, which should nevertheless be sufficient for simulation purposes, since the peak current of the first positive and first negative peak as well as the FWHM of the measured and simulated discharge waveform matches fairly well.

However, the goal of a CDM simulation must be to *predict* a discharge waveform, not only to reproduce it. In order to predict discharge waveforms, much more detailed models for tester, package and parasitics of the die must be developed. Promising approaches for models which are closer to reality are discussed in Narita (1999) and Mergens (2000). The first step is to characterize the test equipment with a well-calibrated metrology chain. A characterization study and the resulting model for a specific commercial CDM tester is presented in (Wilkening, 2001).

The crucial parameter is the arc resistance R_{arc} which to a first approximation is taken as a constant serial resistance. The arc resistance is, a priori, not known. Even in the first approximation, it depends on several environmental and set-up parameters. The physics of the arc which is formed between two approaching bodies is complicated and depends on parameters like the composition and pressure of the gas between the layers, the shape and surface of the bodies, the electrostatic voltage and amount of charge, and the speed of approach (see the discussions in Renninger 1991; Frei 1999; Bönisch 2001). For simulation purposes, the detailed models, which describe the IV characteristic of an arc based on, for example, e.g.,

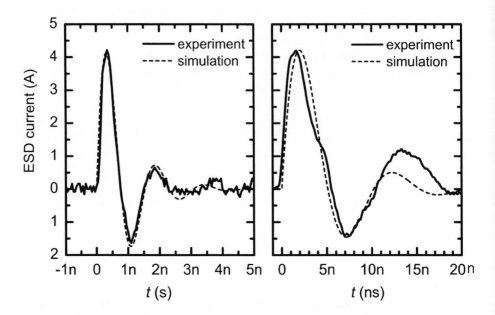

Figure 1.8: Current waveform of a 500 V CDM (left) and a 500 V SDM (right) discharge
and simulation using the model of Figure 1.1. For CDM the experiment is best
fitted by the parameter set $L_\mathrm{s} = 20$ nH, $C_\mathrm{CDM} = 3$ pF, $R_\mathrm{CDM} = 25\ \Omega$. For
SDM $L_\mathrm{s} = 120$ nH, $C_\mathrm{CDM} = 22$ pF, $R_\mathrm{SDM} = 40\ \Omega$ gives a rough estimation.
(Data taken with kind permission from T. Brodbeck.)

Toepler's law or the Rompe–Weisel model and the corresponding refinements and
extensions (for an excellent overview see Renninger 1991), are too complicated
and lead to convergence problems during the CDM simulation of an IC. Hence,
the best approach today is probably to use a constant resistor. $R_\mathrm{arc} = 5$–30 Ω
seems to be a reasonable value.

Finally, it should be noted that in analogy to HBM, CDM exists at the board
level (Charged Board Model, CBM), too. Shaw (1985), Lin (1994), and Maier
(2001) discussed damage to integrated circuits on PCBs which are caused by the
discharge of a charged PCB. In contrast to system level HBM, only very little
work is done on the CBM. If CBM is a concern and should be addressed by
simulation methods, models according to CDM have to be set up, albeit of course
with changed parameter sets and with more focus on distributed parasitics.

1.3   ESD failure mechanisms

Turning now to the effect electrostatic discharges have on the stressed device, a
distinction has to be made between the reversible behaviour and the regime of
irreversible damage of the electronic circuit. As mentioned in the introduction,
even ICs with smallest circuit elements can survive a discharge up to a certain

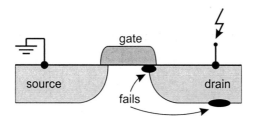

Figure 1.9: Typical location of a *pn* junction-failure caused by ESD.

current level, if protective measures are taken. However, raising the stress level further will inevitably lead to damage to the circuit. The failure modes induced by such damage can be manifold, as discussed in the following.

Since a comprehensive treatment of ESD failure mechanisms and locations is a wide field, going much beyond the focus of this book, only an overview of the major failure mechanisms will be given in this short chapter. For more details Amerasekera (2002) is recommended.

ESD failure mechanisms can be roughly divided into two groups, distinguished by the path taken by the ESD pulse through the device. One group of mechanisms are characterized by failures in the interface circuitry, where one electrode of the failed device is directly connected to an I/O pad and the stressed path is more or less hit by the full power of the ESD pulse.

Typical failures in this circuit part are:

- Local melting of the *pn* junction at the source/drain-to-well border or collector/emitter-to-base. Depending on the doping profile this occurs preferentially at the gate edge or close to the contact holes (Figure 1.9). The discharge path can be either inside the device, e.g., from drain to source, or to neighbouring devices such as a nearby *n* well connected to the VDD supply line.

- Breakdown of the dielectric, i.e. the gate oxide. Because of the higher breakdown voltages of the dielectric this usually occurs in cases where inappropriate circuit protection is chosen, e.g. without any *pn* junction in parallel to the dielectrics. However, due to the aggressive shrink of the oxide thickness in modern sub-100 nm technologies the voltage drop in the protection path has to be carefully controlled to avoid damage to the gate oxide. During a CDM event where large currents appear the resulting high voltage drop across the protective path can easily cause an oxide damage.

The second type of ESD failure occurs if the voltage difference between supply lines is not sufficiently clamped during ESD. Larger potential differences between power supplies like VDD and VSS or VDD and the VDDx of a nearby circuit block will eventually cause damages deep in the core region of an IC. In general, only a small portion of the pulse current follows this route. But the devices in the core which cannot usually rely on specific layout measures to increase their ESD

robustness, tend to become damaged at lower currents. Typical failure locations
are:

- Breakdown between n diffusions connected to different supply domains.

- Triggering of parasitic substrate SCRs between different supply domains.

- Failure of NFETs with small fingers in series to large PFET devices (e.g.
  clock drivers) which can provide enough current to damage the NFET if
  they are transiently switched on during ESD.

- Metal open due to current overload of narrow metal lines. This can occur if
  low ohmic paths between the supplies are triggered and a significant part of
  the current goes across the thin metal wiring.

The essential approach for detecting a failure location is the physical failure analysis, which employs methods such as liquid crystal analysis, emission microscopy
(EMMI), scanning electron microscopy (SEM) and atomic (AFM) or capacitive
(CFM) force microscopy. A good description of these methods is given by Roas
(1999). The conventional approach is to use liquid crystal analysis, which detects
local heating, to pin down the failure to an area no more than several ten square
µm. Alternatively EMMI may be used: this collects the radiation originating from
a leaky junction with a non-ohmic characteristic. Then SEM or even AFM can
be used to investigate the exact nature of the failure, e.g. whether the gate or the
contact holes are damaged. A collection of typical SEM failure pictures is given
in Figure 1.10.

    In rare cases where none of these methods leads to success, the focused ion beam
(FIB) technique can be used to isolate the critical circuit part. By a sequence
of cutting the metal connections and probing on micropads in the investigated
circuit, the failing device can be found. The final step in the failure analysis is to
construct the discharge path. This is done by combining the information about
the stressed contacts, the stress polarization and the failure location. However,
this often leads to some ambiguity, especially if complicated discharge paths across
several devices in series are involved. In this case some modern methods can be
applied to detect the ESD discharge directly during the ESD event. One technique
is backside laser interferometry (BLI), which detects local heating by changes of
the temperature dependent refractive index with a time resolution of less than
10 ns (see Section 1.4.2). This method even provides quantitative information
about the local temperature increase. Another method is based on the standard
EMMI analysis. The detection of the radiation is done here during the discharge
(Russ, 1998; Kessler, 2001). Both methods reveal the elements which are forward
biased respectively driven into breakdown during the ESD pulse. The damage
location itself cannot be deduced from the in-situ measurement, but has to be
found subsequently by the methods described above.

    There is another group of failures which are rather harder to detect. If the
"damage" is manifested by additional traps in a gate oxide, it might cause only
a minor leakage increase and will eventually heal out by high temperature (or
long time) storage. Compared to the previously discussed "hard" failures, the

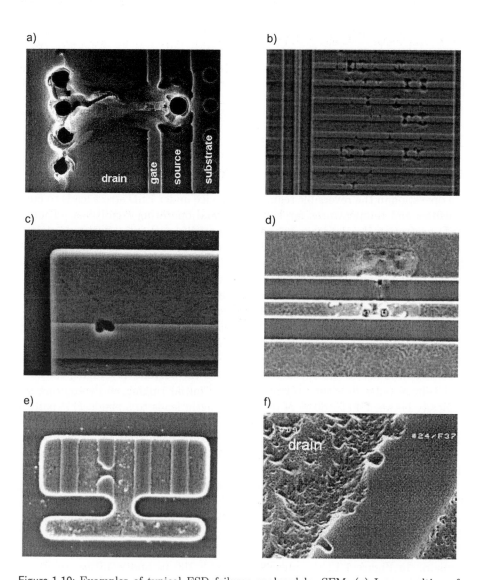

Figure 1.10: Examples of typical ESD failures analysed by SEM. (a) Large melting of
diffusion regions across the gate in a NFET due to filamentation, (b) damage
of PFET driver at several positions, (c) gate oxide damage, (d) parasitic
breakdown between guard rings, (e) failure of small NFET in inverter-type
structure, (f) small filaments at the gate/drain region resulting in small
leakage current which can be annealed.

detection of these "soft" failures requires a well-thought electrical operation to find the failure location. In cases where the failure is not detected by testing the operation of the IC, one speaks of a latent failure. This type of failure causes some concern because of the potential negative influence on reliability (Rainer, 2000). Methods which allow the discharge path to be traced, in combination with preparation techniques like FIB, can be applied if there is the suspicion of such a latency.

1.4   Quantitative characterization methods

1.4.1 IV characterization by square pulsing

An operation in the reversible regime of a device under ESD stress leads to current densities and temperatures far beyond normal operating conditions. Therefore, ESD protection development is almost impossible without a detailed knowledge of the high-current characteristic of protection elements and protected devices. However, parameters of the high-current characteristics relevant to ESD are only accessible by a pulsed measurement, to avoid damage to the device by the IV characterization itself. Square pulse methods are *the* tools for the ESD engineer and indispensable for the design of effective protection concepts.

The basic principle of a square pulse characterization is shown in Figure 1.11. A current pulse with a defined amplitude, here with a duration of 100 ns, is forced into the device. The response of the device is measured, using an oscilloscope, as a voltage drop across the device together with the current. Averaging the current and voltage pulse in a time regime after the initial ringing and triggering spike, typically between 40 %–80 % of the pulse length, leads to a single I/V pair. The complete IV characteristic is obtained by stepping the pulse current amplitude from zero towards the failure current I_{t2}. After each step the low-current IV characteristic is determined by a DC measurement to check the intactness of the device after the preceding square pulse.

The most prominent implementation of square pulsing is Transmission Line Pulsing (TLP). Since the pioneering work of Maloney (1985), TLP has become one of the standard tools for ESD characterization in the HBM regime with pulse durations of about 100 ns. Gieser (1996) developed a very-fast TLP system (vf-TLP) with a pulse width as low as 3.5 ns, to characterize devices in the CDM time domain. In Figure 1.12 the pulse shapes of a 100 ns and a 4.0 ns square pulse generated by a conventional TLP and a vf-TLP system, respectively, are compared with the corresponding HBM and CDM/SDM discharge.

A comprehensive summary of today's TLP systems is given by Gieser (2002). Besides TLP systems, more and more solid-state pulse sources are used for square pulse characterization.

In contrast to HBM, MM or CDM/SDM testing, the focus of square pulsing is clearly on the high-current IV characteristic. Any failure threshold I_{t2} gained by square pulse experiments has to be considered with care. Today it is well established that the failure threshold current is also a function of the rise time of the stress pulse, making a correlation between square pulsing and HBM rather

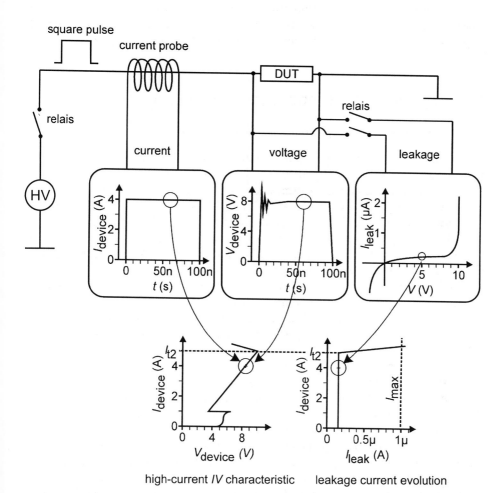

Figure 1.11: Schematic of the square pulsing methodology. In the example, a current pulse with a length of 100 ns is forced into the device. The current through the device (here: $I_{device} = 4$ A) and the voltage across the device (here: $I_{device} = 8$ V) are recorded by a fast oscilloscope. After each pulse stress the DC characteristic is measured and the defined limits are monitored (here: $I_{leak} = 0.25$ μA at the signal voltage $V = 5$ V). By gradually increasing the current of the pulse, a high-current IV characteristic and the corresponding leakage current evolution can be obtained.

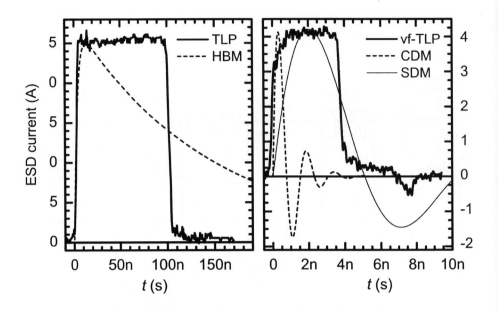

Figure 1.12: Typical current waveforms for TLP and vf-TLP. Left: simulated HBM
(V_{HBM} = 4 kV) and measured TLP (t_{pulse} = 100 ns, I_{TLP} = 2.5 A,
t_{rise} = 1 ns), right: simulated CDM (V_{CDM} = 500 V, C_{CDM} = 10 pF),
simulated SDM (V_{SDM} = 500 V, C_{CDM} = 22 pF) and measured vf-TLP
(t_{pulse} = 4 ns, I_{TLP} = 4 A, t_{rise} = 200 ps).

difficult (Stadler, 1997; Barth, 2001). Currently there are attempts to standardize
the square pulse method in order to get a correlation between the I_{t2} values gained
by different methods and set-up variants (ESDA–TLP, 2002). For vf-TLP the
correlation to CDM is even worse, owing to the fact that vf-TLP represents a two-
terminal test method, while the real CDM is a charge/discharge of a grounded/pre-
charged device through one single pin. Consequently, the current flow in the device
may be totally different for each test method (Gieser, 1999).

Considering device simulation, the great advantage of square pulsing is that
the equivalent circuits are rather simple compared to HBM and CDM. Usually a
simple current source with a piecewise linear current/time behaviour is sufficient.
The steepness of the rising and falling edges of the current pulse is, in good approximation, independent of the DUT and can be extracted from the measurements of
an oscilloscope.

1.4.2   Thermal mapping

Recently, a method has been developed for measuring the local average temperature in the semiconductor during an ESD event (Fürböck, 1999). It is based
on the principle of backside laser interferometry (BLI, see Figure 1.13). In this
method the bulk of the device under investigation is probed by a laser beam from

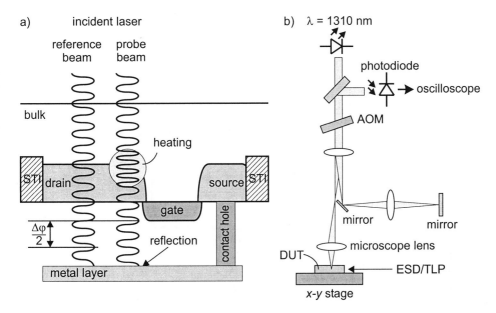

Figure 1.13: (a) Principle of thermal mapping indicating the resulting phase shift between a reference and a probe beam crossing an thermal inhomogeneity. (b) Set-up of a backside laser interferometry according to Fürböck (1999), using an heterodyne beam path where the reference beam path is guided outside of the device under test.

the chip backside. Because of the dependence of the refraction index n on temperature T, the optical path length along the beam changes if the temperature of the semiconductor is increased in the volume probed by the laser beam. The change in the optical path length leads to an optical phase shift $\Delta\varphi$, which is detected interferometrically by interfering the probe beam with a reference beam. Neglecting temperature dependence of the refractive index n (i.e. $\mathrm{d}n/\mathrm{d}T = \mathrm{const.}$) and using the geometric optic approach (Born, 1980) the phase shift can be expressed as

$$\Delta\varphi(x,y,t) = 2 \times \frac{2\pi}{\lambda} \frac{\mathrm{d}n}{\mathrm{d}T} \int_0^d \Delta T(x,y,z,t)\,\mathrm{d}z \qquad (1.9)$$

where λ is the free space wavelength of the laser and d the thickness of the probed layer. An exact expression and its correlation to temperature change, thermal energy, and power dissipating sources can be found in Pogany (2002a) and Pogany (2002c).

This method combines a high temperature sensitivity down to changes of 1 K along an optical path of 3 μm (corresponding to 6 mrad) with an extreme dynamical range which also allows the detection of temperature increase during an ESD event with a local temperature reaching the melting point of silicon at 1685 K. The time evolution of the local temperature integral can be monitored with a proven time resolution < 1 ns (Bychikhin, 2002), as shown in Figure 1.14. To avoid any influence from absorption, IR light below the bandgap of silicon is used

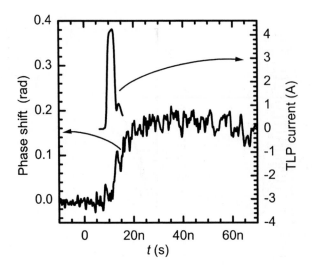

Figure 1.14: Time evolution of the phase shift after a 4 ns vf-TL pulse (Bychikhin, 2002).

(typically $\lambda = 1.3$ μm). Thus, physically the spatial resolution is limited to about $\lambda/(2n) = 0.2$ μm, in practice currently a spatial resolution of 1.5 μm is reported.

The experimental set-up allows scanning of the structure under investigation to build up a complete 2D temperature contour plot (Figure 1.15). However, for this procedure a multiple scanning is required. Any irreversible effects such as caused by stress currents close to the failure threshold, are difficult to investigate by this method. In this case it would be preferable to obtain a 2D image of the phase shift using just one stress pulse. This can be achieved by a broad beam illumination using holographic interferometry (Pogany, 2002b). The thermal image is obtained during one time window (5 ns long) during the stress pulse.

Another feature of the thermal mapping method is the occurrence of a non-thermal, free charge carrier related effect influencing the phase shift (the plasma-optical effect). The change in carrier density, arising from the large current density during the ESD discharge, modifies the refractive index, too (Fürböck, 1999). The effect is weaker than the temperature related effect and causes a phase shift with the opposite sign. Both effects can be resolved in the experiment, for example by measuring the temporal development of the phase shift (Figure 1.16). As an example the phase shift analysis is carried out at two different positions along the device length of a smart power transistor (Fürböck, 2000). The position A corresponds to a hotspot at a pn junction which is reverse biased and operating under electrical breakdown conditions during ESD. Opposed to that the position B is next to a current injection position (emitter–base injection) of a forward biased pn junction. During the ESD pulse the hotspot A shows a positive phase shift as a result of the increasing temperature and cools down after the end of the pulse. The area around the forwarded pn junction shows first a negative phase shift signal as a result of current injection (plasma-optical effect). Later it turns to a positive phase shift because of the heat spreading from the hotspot A into the area around

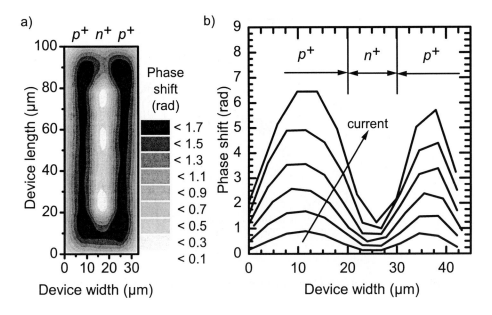

Figure 1.15: a) 2D plot of the phase shift for a vertical *npn* at the end of a 150 ns TL pulse ($I_{TLP} = 0.4$ A). b) Phase shift profiles for different currents along a cut line in the centre of the device length at 50 µm (Fürböck, 1999).

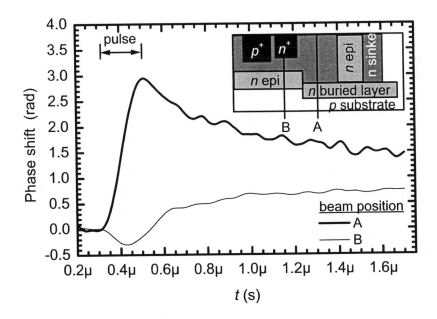

Figure 1.16: Phase shift evolution measured for two positions in a test structure stressed by a 1 A pulse of 150 ns duration (Fürböck, 2000).

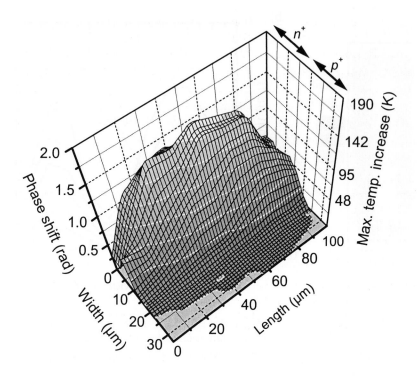

Figure 1.17: Phase shift and calculated maximum temperature as function of position (Fürböck, 1999).

location B.

The BLI method described above provides the temperature integral within the total volume probed by the laser beam. However, by combining BLI with device simulation, a profile of the temperature distribution can be built up, as shown in Figure 1.17. The crucial advantage of this approach in comparison to other optical detection methods used for ESD investigations, like emission microscopy (Kessler, 2001), is the availability of time resolved, quantitative temperature data. This allows verification of the results of the device simulation and, if required, improvement of the models and calibration of the parameters in the high temperature regime.

Summary

- Today, three different models are used to replicate "real world" component ESD events, namely HBM, MM, and CDM/SDM. The discharge of these models can be, to a first approximation, modelled by an RLC network.

  - HBM is a discharge of a 100 pF capacitance into a 1.5 kΩ resistor.

The over-damped waveform of an HBM event has a peak current of ≈ 1.33 A for 2 kV discharge voltage, a rise time of ≈ 8 ns and a FWHM of ≈ 100 ns. To model the HBM waveform, the test board capacitance and the stray capacitance must be included in the equivalent circuit.

- MM represents a discharge of a 200 pF capacitance into a low impedance (nominal 0 Ω). The waveform depends on the (parasitic) impedance and the inductance. Typical waveforms have a peak current of ≈ 2.6 A for 200 V discharge voltage, a rise time of ≈ 10 ns and a FWHM of ≈ 25 ns. The discharge can be well described by a simple RLC model.

- CDM and SDM model the charge or discharge of a grounded or charged component. The waveform of CDM is mainly governed by the capacitance of the device and the arc resistance. Waveforms typically show an amplitude of 5–10 A for a 500 V discharge, a rise time of 200–400 ps, and a FWHM of ≈ 1 ns (SDM 3 ns). CDM can be modelled by an equivalent circuit consisting of R, L, C; however, the arc resistance is not easily predictable.

• Square pulsing, and particularly TLP and vf-TLP, are *the* standard analysis tools for characterizing the device behaviour, particularly turn-on and high-current characteristics. For simulation purposes, the current waveforms can be approximated by a piece-wise linear current source.

• Recently, optical mapping (BLI) has been proven to be a very powerful analysis method for investigating device behaviour under ESD stress on a microscopic scale. BLI is the only experimental method which can yield information about the crucial parameters such as carrier and temperature distribution in-situ during an ESD event, on a nanosecond time scale and with a μm spatial resolution.

Bibliography

Amerasekera A., Verweij J., "ESD in Integrated Circuits", Quality and Reliabilty Eng. Int., **8** (1992), 259.

Amerasekera A., Duvvury C., *ESD in Silicon Integrated Circuits, Second Edition,* John Wiley, Chichester, England, 2002.

Antinone R.J., *Electrical Overstress Protection for Electronic Devices,* Noyes Publications, New Jersey, 1986.

Barth J., Richner J., Verhaege K., Henry L.G., "TLP-Calibration, Correlation, Standards, and New Techniques", Proc. 22nd EOS/ESD Symposium (2000), 85.

Barth J., Richner J., "Correlation Considerations: Real HBM to TLP and HBM Testers", Proc. 23nd EOS/ESD Symposium (2001), 453.

Baumgärtner H., Gärtner R., *ESD, Elektrostatische Entladungen (in German),* Oldenbourg, Munich, Germany, 1997.

Beebe S., "Simulation of Complete CMOS I/O Circuit Response to CDM Stress", Proc. 20th EOS/ESD Symposium (1998), 259.

Bhar T.N., McMahon E.J., *Electrostatic Discharge Control,* Hayden, NJ, 1983.

Bönisch S., Pommerenke D., Kalkner W., "Broadband Measurement of ESD Risetimes to Distinguish Between Different Discharge Mechanisms", Proc. 23rd EOS/ESD Symposium (2001), 373.

Born M., Wolf E., *Principle of Optics,* Cambridge University Press, England, 1980.

Brodbeck T., Kagerer A., "Influence of the Device Package on the Results of CDM Tests – Consequences for Tester Characterization and Test Procedure", Proc. 20th EOS/ESD Symposium (1998), 320.

Bychikhin S., Dubec V., Litzenberger M., Pogany D., Gornik E., Groos G., Esmark K., Stecher M., Stadler W., Gieser H., Wolf H., "Investigation of ESD Protection Elements under High Current Stress in CDM-like Time Domain Using Backside Laser Interferometry", Proc. 24th EOS/ESD Symposium (2002), 387.

Corp M.B., *Zzaap! Tarning ESD, RFI, and EMI,* Academic, London, 1990.

Dabral S., Maloney T.J., *Basic ESD and I/O Design,* John Wiley, New York, 1998.

Dangelmayer T., *ESD Program Management; A Realistic Approach to Continuous Measurable Improvement in Static Control,* Van Nostrand Reinhold, New York, 1990.

Diaz C.H., Kang S.-M., Duvvury C., *Modeling of Electrical Overstress in Integrated Circuits,* Kluwer International, Boston, Mass., 1995.

EN 61000–4–2, Part 4, Section 2: Electrostatic Discharge Immunity Test, European Standard, 1996.

ESD Association, *ESD Standard Test Method for Electrostatic Discharge Sensitivity Testing — Charged Device Model (CDM) Component Level (ESD STM5.3.1–1999),* of ESD Association Working Group WG 5.3.1, 2001.

ESD Association, *ESD Standard Test Method for Electrostatic Discharge Sensitivity Testing — Human Body Model (HBM) Component Level (ESD STM5.1–2001),* of ESD Association Working Group WG 5.1, 2001.

ESD Association, *ESD Standard Test Method for Electrostatic Discharge Sensitivity Testing — Machine Model (MM) Component Level (ESD STM5.2–1998),* of ESD Association Working Group WG 5.2, 1998.

ESD Association, *Socket Device Model (SDM) Tester, an application and technical report (ESD TR 08–00),* of ESD Association Working Group WG 5.3.2, 2000.

ESD Association, *Standard Practice for Electrostatic Discharge Sensitivity Testing — Transmission Line Pulse (TLP) Component Level,* of ESD Association Working Group WG 5.5, work started.

Frei S., Senghaas M., Jobava R., Kalkner W., "The Influence of Speed of Approach and Humidity on the Intensity of ESD", Proc. EMC Symp. (1999), 105.

Fürböck C., Pogany D., Litzenberger M., Gornik E., Seliger N., Gossner H., Müller-Lynch T., Stecher M., Werner W. "Interferometric Temperature Mapping during ESD Stress and Failure Analysis of Smart Power Technology ESD Protection Devices", Proc. 21st EOS/ESD Symp. (1999), 241.

Fürböck C., Esmark K., Litzenberger M., Pogany D., Groos G., Zelsacher R., Stecher M., Gornik E., "Thermal and Free Carrier Concentration Mapping during ESD Event in Smart Power ESD Protection Devices Using an Improved Laser Interferometric Technique", Microelectronics Reliability **40** (2000), 1365.

Gärtner R., "Field-induced CDM-Ausfall in der Halbleiterfertigung *(German only)*", Proc. 6th ESD-Forum (1999), 49.

Gieser H., Haunschild M., "Very-Fast Transmission Line Pulsing of Integrated Structures and the Charged Device Model", Proc. 18th EOS/ESD Symposium (1996), 85.

Gieser H., *Methods for the Characterisation of Integrated Circuits Employing Very Fast High-Current Impulses (in German),* PhD thesis, Technical University Munich, 1999; published Shaker-Verlag, Aachen, Germany, 1999.

Gieser H., in *ESD in Silicon Integrated Circuits,* ed. Amerasekera A., and Duvvury C., John Wiley, Chichester, England, 2002.

Greason W.D., *Electrostatic Discharge in Electronics,* Research Studies Press, Taunton, England, 1992.

IEC, *Electrostatics, Part 3–1: Methods for Simulation of Electrostatic Effects – Human Body Model (HBM) – Component Testing,* 61340–3–1:2002.

IEC, *Electrostatics, Part 3–1: Methods for Simulation of Electrostatic Effects – Machine Model (MM) – Component Testing,* 61340–3–2:2002.

JEDEC Solid State Technology Association, *Electrostatic Discharge (ESD) Sensitivity Testing Machine Model (MM),* JESD22–A114–A, 1998.

JEDEC Solid State Technology Association, *Electrostatic Discharge (ESD) Sensitivity Testing Human Body Model (HBM),* JESD22–A114–B, 2000.

JEDEC Solid State Technology Association, *Field-Induced Charged Device Model Test Method for Electrostatic Discharge Withstand Thresholds for Microelectronic Components,* JESD22–C101–A, 2000.

Kessler T., Kederer F., Wulfert F.W., "LiveZap – A During-the-Event Damage Site Localization Method", Proc. 7th ESD-Forum (2001), 71.

Lin D.L., Jon M.-C., "Off-chip Protection: Shunting of ESD Current by Metal Fingers on Integrated Circuits and Printed Circuit Boards", Proc. 16th EOS/ESD Symposium (1994), 279.

Maier R., "ESD aus Sicht eines Kfz-Herstellers *(German only)*", Proc. 7th ESD-Forum (2001), 49.

Maloney T., Khurana N., "Transmission Line Pulsing Techniques for Circuit Modeling of ESD Phenomena", Proc. 8th EOS/ESD Symposium (1985), 49.

Matisoff B.S., *Handbook of Electrostatic Discharge Control*, Van Nostrand Reinhold, New York, 1986.

Mergens M., Wilkening W., Kiesewetter G., Mettler S., Wolf H., Hieber J., Fichtner W., "ESD-level Circuit Simulation – Impact of Gate RC-Delay on HBM and CDM Robustness", Proc. 22nd EOS/ESD Symposium (2000), 446.

MIL STD 883.C/3015.7 notice 8, in *Military Standard for Test Methods and Procedures for Microelectronics: ESD Sensitivity Classification*, March 22, 1989.

Mußhoff C., Wolf H., Egger P., Gieser H., Guggenmos X., "Risetime Effects of HBM and Square Pulses on the Failure Thresholds of GGNMOS Transistors", Proc. ESREF (1996), 1746.

Narita K., Horiguchi Y., Hayano K., Suzuki K., "A Simulation Analysis of Quarter-Micron CMOS LSI Input Circuit Behaviour under CDM-ESD for Protection Device Improvement", Proc. 21st EOS/ESD Symposium (1999), 116.

Pogany D., Bychikhin S., Litzenberger M., Gornik E., Groos G., Stecher M., "Extraction of Spatio-temporal Distribution of Power Dissipation in Semiconductor Devices Using Nanosecond Interferometric Mapping Technique", Appl. Phys. Lett. **81** (2002a), 2881.

Pogany D., Dubec V., Bychikhin S., Fürböck C., Litzenberger M., Groos G., Stecher M., Gornik E., "Single-Shot Thermal Energy Mapping of Semiconductor Devices With the Nanosecond Resolution Using Holographic Interferometry", IEEE Electr. Dev. Lett. **23** (2002b), 606.

Pogany D., Bychikhin S., Fürböck C., Litzenberger M., Gornik E., Groos G., Esmark K., Stecher M., "Quantitative Internal Thermal Energy Mapping of Semiconductor Devices under Short Current Stress Using Backside Laser Interferometry", IEEE Trans. Electr. Dev. **49** (2002c), 2070.

Rainer J.C., Keller T., Jäggi H., Mira S., "Impact of ESD-induced Soft Drain Junction Damage on CMOS Product Lifetime", Microelectr. Reliability **40**, (2000), 1649.

Renninger R.G., Jon M., Lin D.L., Diep T., Welsher T.L., "A Field-Induced Charged Device Model Simulator", Proc. 11th EOS/ESD Symposium (1989), 59.

Renninger R.G., "Mechanisms of Charged Device Model Electrostatic Discharges", Proc. 13th EOS/ESD Symposium (1991), 127.

Roas R., Boit C., Staab S. (editors) *Microelectronic Failure Analysis,* Proceeding of the ASM International, Ohio, 1999.

Roozendaal van L.J., Amerasekera A., Bos P., Baelde W., Bontekoe F., Kersten P., Korma E., Rommers P., Krys P., Weber U., Ashby P., "Standard ESD Testing", Proc. 12th EOS/ESD Symposium (1990), 119.

Russ C., Gieser H., Verhaege K., "ESD Protection Elements During HBM-ESD-stress tests – Further Numerical and Experimental Results", Proc. 16th EOS/ESD Symposium (1994), 96.

Russ C., Verhaege K., Bock K., Roussel P.J., Groeseneken G., Maes H.E., "A Compact Model for the Grounded-Gate nMOS Behaviour under CDM ESD Stress", Proc. 18th EOS/ESD Symposium (1996), 302.

Russ C., Bock K.-H., Rasras M., de Wolf I., Groeseneken G., Maes H.E., "Non-Uniform Triggering of gg-nMOSt Investigated by Combined Emission Microscopy and Transmission Line Pulsing", Proc. 20th EOS/ESD Symposium (1998), 177.

Shaw R.N., Enoch R.D., "An Experimental Investigation of ESD-Induced Damage to Integrated Circuits on Printed Circuit Boards", Proc. 7th EOS/ESD Symposium (1985), 132.

Stadler W., Guggenmos X., Egger P., Gieser H., Mußhoff C., "Does the ESD Failure Current Obtained by Transmission Line Pulsing Always Correlate to Human Body Model Tests?", Proc. 19th EOS/ESD Symposium (1997), 366.

Verhaege K., Roussel P.J., Groeseneken G., Maes H.E., Gieser H., Russ C., Egger P., Guggenmos X., Kuper F.G., "Analysis of HBM ESD testers and Specifications Using 4th Order Lumped Element Model", Proc. 15th EOS/ESD Symposium (1993), 129.

Wang A.Z.H., On-Chip ESD Protection for Integrated Circuits, Kluwer Academic Publishers, Boston, 2002.

Wilkening W., Stadler W., Willemen J., Esmark K., Wolf H., Gieser H., "Investigations on Charged Device Model for On-Chip ESD Protection in the ASDESE Project", Proc. 7th ESD-Forum (2001), 87.

Yan K.P., Gärtner R., Lim S. "A Study of Wafer Level ESD Testing", Proc. 23rd EOS/ESD Symposium (2001), .

Yokoi K., Watanabe T., "An Effective ESD Protection System in the Back End (BE) Semiconductor Manufacturing Facility", Proc. 22th EOS/ESD Symposium (2000), 1.

BIBLIOGRAPHY

Chapter 2

Parameters for an ESD protection concept

2.1 Introduction

An ESD stress forces a high current pulse (1–15 A) with a short duration (typically about 3–100 ns) and a short rise time (100 ps–10 ns) between two pads (here "pad" denotes the metallic interface of the integrated circuit (IC) to the "outside world"). In the real world, a pulse may stress any pad with respect to any other pad. Consequently, all elements between any pad combination must be either capable of handling that current, or else appropriate ESD protection elements must shunt the current between the pads. From an ESD engineer's point of view, any IC may be reduced to simple circuits and elements as depicted in Figure 2.1.

- Output cells drive a signal to the pad. Typically inverter stages consisting of NFET and PFET devices are connected with the drain diffusion to the pad.

- Input cells receive signals from external devices. In this I/O cell type, gates of the inverter stages are usually connected to the pad.

- The "core" circuit mainly provides the functionality of the IC. Typical core blocks are memories, digital logic, but also complex analog macros such as ADC/DAC (analog to digital converter/digital to analog converters). A common definition of a core block is that no element of the core has a direct, low-ohmic connection to a pad of an input/output (I/O) cell.

- Different ground and power cells provide the power supply for the core blocks and the signal cells. In contrast to signal cells, the voltage of the power supply is fixed during operation. A modern IC typically contains several different power supply domains. In Figure 2.1 there are three different supply domains depicted (VDD1/VSS1 as supply for the output pad, VDD2/VSS2 as supply for the input pad, VDD/VSS as core supply).

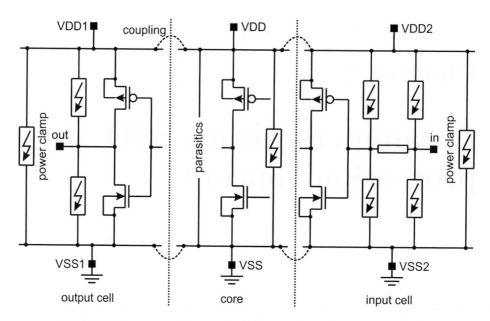

Figure 2.1: Representation of an IC containing output and input cells and core logic with the most important elements which have to be considered for the ESD development. ESD shunt elements of a possible protection concept are included.

- Both I/O cells and the core logic contain so-called "parasitic elements" ("parasitics"). Such parasitics could be, for example, bipolars formed by two diffusions tied to different potentials, which are close together (e.g., an n^+ guard ring and an n^+ active area). These parasitics are at high risk, as they may experience much higher voltage differences during ESD than under normal operation.

For all circuit types, power supply arrangements, and parasitics mentioned above, adjusted ESD measures must be defined. The different considerations which must be taken into account for this will be discussed in the following sections.

2.2　Basic ESD protection concepts

The fundamental idea of on-chip ESD protection is to provide the ESD pulse with a robust discharge path that is not destroyed by the high current. This path has to be considered to its full extent, from the point where the pulse is forced into the IC to the contact where the pulse "leaves" the chip to the grounded electrode. This has to be ensured for any possible combination of pins involved in an ESD discharge. An example for such a protection path concept is given in Figure 2.2.

In a good ESD protection concept every part of every possible discharge path is well controlled. Any residual randomness might later on lead to failures in the product. Essential parts of the discharge path are the I/O circuit, the power

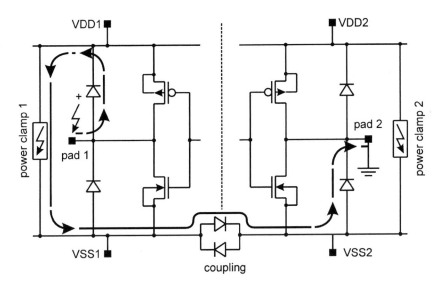

Figure 2.2: Example of a typical ESD shunt path between two output pads using diodes and power clamps as primary protection elements. Pad 1 is stressed, pad 2 grounded.

bus resistance, the power clamp as protection between the different supplies, and the coupling elements between different power supply domains. Both for the I/O circuit and the power clamps, there is the option either to use the existing active circuitry (sufficient ESD robustness provided) or to add a parallel protection path with specific ESD protection devices to shunt the ESD current. The use of elements from the active I/O circuitry as drivers in the main ESD discharge path is also called self-protection. In order to achieve the required robustness, design and layout measures have to be implemented which might not be compliant with the required performance features under normal operation of the circuit. In this case it is better to install a parallel ESD protection element and to take care that most of the ESD discharge current flows through this protection element.

ESD specific elements of the circuitry can be classified as shunt elements or current limiting devices (Figure 2.3). The latter are mostly resistors which reduce the current for unwanted discharge paths (such as into the circuit). For I/O cells with pure ac signalling, blocking capacitors can also be used.

The shunt elements must provide a low resistance, robust path in ESD conditions, to avoid extreme overvoltages which might damage the protected I/O circuit, such as gate oxides. Moreover, the major part of the ESD discharge current should be conducted through the protection element to relieve the stress on the output drivers (Figure 2.4). At the same time they should not degrade the normal operation, so the breakdown voltage and leakage must be properly adjusted. Thus, the operation of the shunt element has to be designed for a narrow voltage window between the normal operation of the I/O circuitry, where the shunt should not be apparent, and the critical, destructive voltages, e.g. oxide breakdown voltage, of

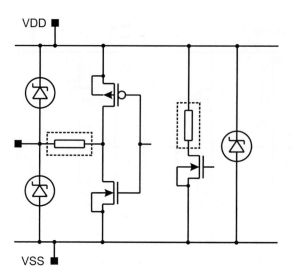

Figure 2.3: Typical ESD protection elements which either limit the current (indicated by dashed boxes) or shunt the current (indicated by circles).

the protected circuit (Figure 2.5). This window is called the ESD design window.

   The creative work of ESD engineers to fulfil the constraints of ESD robustness, low on-resistance, minimized performance distraction and minimized area consumption has led to the development of a large number of ESD protection devices during the past three decades. Table 2.1 gives an overview of these devices and their preferred fields of application.

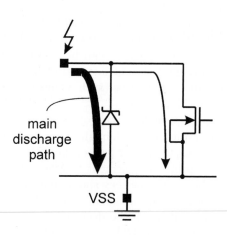

Figure 2.4: Schematic output ESD protection.

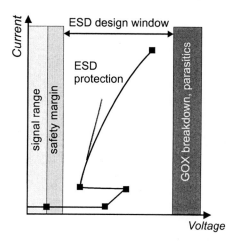

Figure 2.5: ESD design window for a protection element exhibiting a so-called snapback in the *IV* characteristic.

Table 2.1: Compilation of commonly used protection elements with their fields of application and major benefits and drawbacks.

Device	Field of application	Rating
grounded-gate NFET	protection element for I/O cells and power clamp; used for all CMOS technologies down to 100 nm node	+ technology robust + flexible concerning circuit requirements − high trigger voltage − area consumption depending on the ballasting resistance
gate-coupled NFET	mainly power clamp, but also protection element for I/O cells; used for all CMOS technologies down to 100 nm node	+ technology robust + low trigger voltage − danger of turn-on because of high-frequency noise − applicable for limited signal frequency regime − area consumption, if only MOS effect is used as shunt path

Table 2.1: continued

Device	Field of application	Rating
lateral bipolar (FOX) transistor	protection elements for I/O cells and power clamp; used for CMOS technologies down to 0.5 µm node	+ technology robust + flexible concerning circuit requirements − high trigger voltage − area consumption depending on the ballasting resistance − not applicable for modern technologies with STI
vertical bipolar transistor	protection element for I/O cells and power clamp; used for bipolar and BiCMOS technologies	+ technology robust + flexible concerning circuit requirements + area efficient − high trigger voltage − possibly latch-up critical
NFET, bipolar, SCR with control circuit	protection element for I/O cells and power clamp; used for all CMOS technologies down to 100 nm node	+ flexible concerning circuit requirements + area efficient + low trigger voltage − delicate adjustment of design/circuit parameters
diode	protection element for I/O cells and coupling device; used for all process technologies	+ in general technology robust + area efficient + good clamping − limited range of circuit application (e.g., no over-voltage tolerance supported) − worse voltage clamping due to power clamp device in series − strong influence of bus resistance − protection for reverse bias required

Table 2.1: continued

Device	Field of application	Rating
diode strings	protection element for I/O cells and power clamp; used for all process technologies	+ appropriate solution for over-voltage tolerance, if GOX devices not applicable − leakage at high temperatures − area consuming − worse voltage clamping due to power clamp device in series − strong influence of bus resistance − protection for reverse bias required
silicon controlled rectifier (SCR)	protection element for I/O cells and power clamp; used for all process technologies	+ very good voltage clamping + area efficient + flexible concerning circuit requirements − latch-up critical − limited technology robustness − often requires parallel diode for reverse polarization − possibly high trigger voltage (often trigger element required)

2.3 Self-protection of FETs

The well-known, straightforward ESD protection concepts of CMOS circuits rely on the self-protection capability of FETs. In I/O cells containing PFET and NFET drivers of large width, the drivers themselves will guarantee a certain ESD hardness even in the absence of further protection measures, since their pn junctions experience a breakdown or a forward biasing during the ESD pulse and shunt the current to one of the supply buses. This provides protection for other more sensitive circuit parts like the gate oxide. If the width of the active driver is insufficient to sustain the required ESD level or a pure input pad is used, FET transistors can be added as passive elements in which the gates are tied to the supplies to switch off the transistor under normal operation conditions. The grounded-gate NFET (ggNFET) is often used for this purpose. An example of a protection using the ggNFET is given in Figure 2.6. In the case of junction breakdown in a driver or a passive element like the ggNFET, the underlying lateral bipolar transistor, the npn of the NFET, or alternatively the pnp of the PFET, will dominate the IV characteristic at high current densities. The lateral npn transistors exhibit a

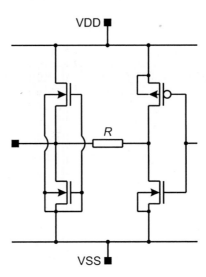

Figure 2.6: ESD protection concept based on a ggNFET. A decoupling resistor R is used in cases where the current across the driver stage has to be limited.

characteristic feature of many elements operated under ESD conditions, namely the snapback of the IV curve. This is discussed for the example of the NFET.

When, after the breakdown of the n diffusion to p well junction (V_{bd}) at the drain side, further current is forced through the structure, the voltage across the device clearly increases (Figure 2.7). However, at the point where the current in the p well leads to a significant forward bias of the source junction, the lateral npn is triggered. The voltage value of the triggering is named V_{t1}. Because of the current gain of this transistor the voltage across the device drops significantly. The value of the lowest voltage after the snapback is commonly referred to as the "sustaining" or "holding" voltage V_h. A negative differential resistance branch is observed in the IV characteristic. After this snapback the voltage steadily increases with growing current with a differential resistance R_{diff} until the so-called second breakdown is reached. Here, a usually irreversible destruction occurs as a result of a thermal overload in a very limited volume of the device. Melted filaments form very low-ohmic paths in the device. The current level at which this effect occurs is noted as I_{t2}, in analogy to the current I_{t1} where the "first", non-destructive snapback appears.

Unfortunately, device size is not the only essential parameter for a high ESD robustness. The mentioned parameters have to be adjusted delicately to get a homogeneous current flow and maximum ESD hardness. This can easily be seen by considering an NFET structure which is fractured into several separated fingers (Figure 2.8). As soon as one of the fingers has triggered at V_{t1}, the voltage drops and this finger takes over the whole current. Usually, owing to minute differences in the doping profile or the gate control, not all fingers trigger at exactly the same voltage value. If the current in the triggered finger exceeds the critical value for destruction before the voltage drop across the device has reached V_{t1} again, the

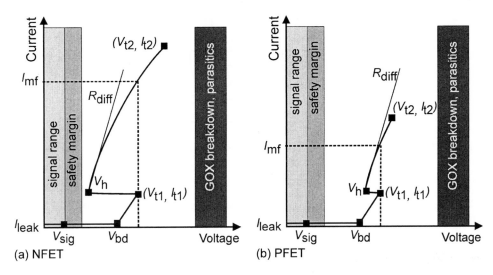

Figure 2.7: High-current *IV* characteristics of (a) NFET and (b) PFET devices. The parameters shown are relevant for the definition of an appropriate ESD protection concept. The drafts are not to scale. In general, the intrinsic ESD robustness of the PFET is worse than the NFET due to the lower snap-back with a higher resulting dissipated power.

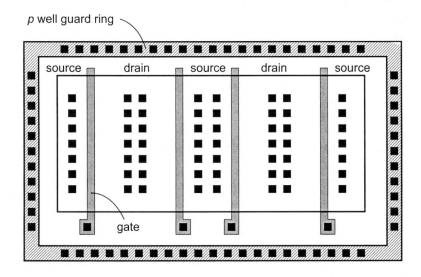

Figure 2.8: Layout draft of a typical multi-finger NFET, as commonly used as driver in output pads.

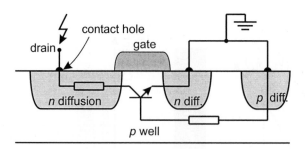

Figure 2.9: Cross section of an NFET with parasitic bipolar transistor and relevant resistances.

ESD hardness is determined by the size of the single finger and not by the whole device.

One method of solving this problem is the concept of ballasting resistance. By introducing a certain resistance in series to each finger it can be arranged that V_{t1} is exceeded at a (multi-finger triggering) current I_{mf} before the destructive current density I_{t2} is reached. This is often realized by increasing the contact-to-gate spacing of the FET, which results in a serial resistance for a breakdown at the gate edge arising from the diffusion resistance between the contact hole and the gate edge (Figure 2.9). However, this adds a significant area to the driver. In technologies with silicided diffusions of extremely low sheet resistance, this concept requires a blocking of the silicide formation between the contacts and the gate edge to achieve an appropriate ballasting resistance. However, this again leads to a performance loss owing to the higher on-resistance of the driver and a higher complexity in the process technology.

Another factor which strongly influences the uniformity of the current flow is the arrangement of the well contacts. For example, the current uniformity can be improved by the placement of well contacts at appropriate positions to allow an interaction between fingers by picking up the local potential of the well. Conversely, any well contacts which are connected to the supply via a low ohmic path, tend to inhibit the triggering.

The most important parameters of an NFET in the high-current regime are summarized in Table 2.2.

In a supply domain "x", the parasitic p^+n diode in the PFET and the n^+p diode in the NFET are commonly used to shunt positive ESD stress to VDDx (high supply potential) and negative ESD stress to VSSx (low supply potential), respectively. If the width of the junction is sufficient and the wells are connected directly to VDDx and VSSx (not the case in over-voltage tolerant pads, I²C pads, and so on), no further ESD protection measures are needed for these stress conditions. Therefore, an analysis of the forward-biased parasitic diode characteristic is also required (Section 2.4.1).

Stacked FETs are commonly used in applications where, owing to reliability issues, the use of a single NFET/PFET between pad and ground potential is not allowed. Typical examples are the 5 V tolerant pads in technologies which only

Table 2.2: Parameters of the NFET in the high-current regime. The parameters of the forward biased n^+p (NFET) and p^+n diode (PFET) are listed in Table 2.3

Parameter	Explanation
I_{leak}	leakage current at maximum operating voltage and temperature
V_{t1}	turn-on voltage of parasitic bipolar transistor
I_{t1}	trigger current of parasitic bipolar transistor
V_h	holding voltage
R_{diff}	differential resistance in the low ohmic state of the device
I_{mf}	lowest current which results in homogeneous multi-finger turn-on of multi-finger structures
I_{t2}, V_{t2}	current level that causes destruction of the protection element under square pulse conditions, and corresponding voltage
V_{clamp}	(clamping) voltage at I_{ESD}

allow a maximal voltage across the gate of 3.3 V. In comparison with single FET devices, stacked FETs have a much larger area consumption and a rather low ESD robustness if no special design or process measures are implemented. One reason for the low ESD robustness is the high holding voltage which leads to a high power dissipation in the device. While the high trigger and holding voltages degrade the self-protection capability, they offer at the same time the choice of a large variety of possible ESD protection elements which can protect these stacked devices.

2.4 Characteristics of protection devices

2.4.1 Diodes

Under certain circumstances, diodes operating under forward bias condition during ESD are the first choice as protection devices. Implemented as standard devices with diffusion in the corresponding well (Figure 2.10) they combine high area efficiency, high ESD and process robustness, and often good clamping features. Their limitation is their unipolar use. If the diode is forced into breakdown, it usually fails at very low ESD levels.

Single diodes can be used in the ESD protection concept in two ways. First, they can be used to shunt the ESD current in a forward biased mode. Examples are a p^+n diode with the p^+ diffusion directed to pad and the n well connected to VDD, or an n^+p diode with the n^+ diffusion on the pad and the p well (substrate/bulk) on VSS. Secondly, a pair of anti-parallel diodes is often used as a coupling device between different power supply blocks, such as between digital VSSx and analog VSSA. In contrast to the FET, the diode (without guard rings)

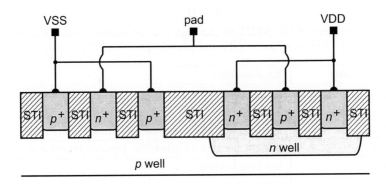

Figure 2.10: Cross section of an n^+p (left) and a p^+n diode (right) used in an ESD
protection concept.

does not show a snap-back behaviour (Figure 2.11). The voltage clamping under
reverse biased condition is very bad and the withstand current I_{t2} is very low.
Consequently, a diode must certainly have an additional element which protects it
under reverse biased stress condition. Therefore, a power supply clamp with sufficient voltage clamping capability is mandatory (Figure 2.12). In this situation,
the added voltage drop across the power supply clamp, the forward biased diode
and the bus resistance must be lower than the V_{t2} of the diode in reverse biased
mode.

The high-current characteristics of p^+n and n^+p diodes are illustrated in Fig-

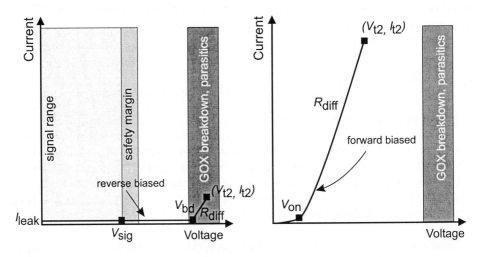

Figure 2.11: High-current IV characteristic of p^+n and n^+p diodes. In general, diodes
can be used as ESD protection elements only in forward biased conditions.
In reverse biased condition the breakdown must be prevented by appropriate protection strategies due to the very low current withstand capability.
Therefore, the breakdown of the p^+n and n^+p junctions are parameters
which define the protection window.

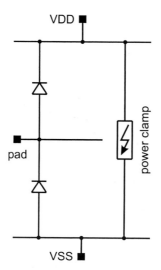

Figure 2.12: ESD protection concept using diodes. Owings to the unipolar use of diodes, an appropriate power clamp is required to avoid stressing the diode in the reverse biased direction.

ure 2.11. The parameters of the high-current characteristic which determine the behaviour of the diodes under ESD stress and the dependence on simple layout parameters are listed in Table 2.3.

An alternative to the junction diode is the FET diode. There, the principle is to conduct the ESD discharge current as FET current of, for example, an NFET. If the maximum current of one to several amps is fed through the device, a very large width is required.

2.4.2   Diode strings

Diode strings show a very similar behaviour to single diodes. Like the single diode, the diode string has a very low withstand current I_{t2} in reverse mode and may only be used as ESD shunt element under forward bias. The main difference is that the on-voltage V_{on} is adjustable by the number of serial diodes with isolated wells (Voldman, 1994). Diode strings can for example be used in over-voltage tolerant pads. The number of serial diodes n which must be used for a given maximum voltage difference between signal voltage V_{sig} and the power supply voltage VDD can be estimated in first order by

$$n > \frac{V_{sig} - \text{VDD}}{V_{on}} \qquad (2.1)$$

where V_{on} is the on-voltage of a single diode.

However, the diodes must not be considered as independent devices. Between p^+ diffusion, n well, and p substrate a parasitic pnp transistor is formed. These parasitic bipolar transistors of the diodes of the string form a Darlington stage

Table 2.3: Parameters of the p^+n diode (n^+p diode in analogy) in the high-current regime for forward and reverse biased mode.

Parameter	Explanation
	forward biased
V_{on}	on-voltage at defined current/(area pn junction)
R_{diff}	differential resistance in the low ohmic state of the device
I_{t2}, V_{t2}	current level that causes destruction of the protection element and corresponding voltage
V_{clamp}	(clamping) voltage at I_{ESD}
	reverse biased
I_{leak}	leakage current at maximum operating voltage and temperature
V_{bd}	breakdown voltage at defined current/(area pn junction)
I_{t2}, V_{t2}	current level that causes destruction of the protection element and corresponding voltage

which amplifies the leakage current of the diode connected to the pad (Figure 2.13). Particularly at higher temperatures, the leakage current might be significantly increased due to the Darlington stage configuration. To reduce the leakage current of diode strings, either snubber diodes (Voldman, 1995) or parallel bias networks (Maloney, 1995) have been proposed.

A characterization of the parasitic vertical pnp bipolar transistor is essential, for reasons both of leakage and of ESD robustness. The high voltage at the diffusion of the diode connected directly to the pad can even cause a breakdown of this parasitic pnp transistor leading to low-level ESD failures. The parameters of the high-current characteristic which determine the behaviour of an n-fold diode string under ESD stress are listed in Table 2.4. In addition, for an optimized diode string

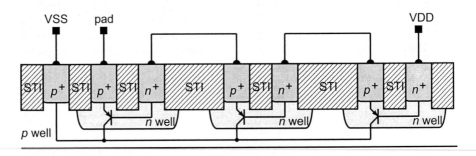

Figure 2.13: Darlington stage formed by the parasitic pnp transistors of stacked p^+n diodes in a common substrate.

Table 2.4: Parameters of an n-fold p^+n diode string (n^+p diode string in analogy) in the high-current regime for forward and reverse biased mode.

Parameter	Explanation
	forward biased
I_{leak}	leakage current at maximum operating voltage and temperature
V_{on}	on-voltage
R_{diff}	differential resistance in the low ohmic state of the device
I_{t2}, V_{t2}	current level that causes destruction of the protection element and corresponding voltage
V_{clamp}	(clamping) voltage at I_{ESD}
	reverse biased
I_{leak}	leakage current at maximum operating voltage and temperature
V_{bd}	breakdown voltage at defined current/(area pn junction)
I_{t2}, V_{t2}	current level that causes destruction of the protection element and corresponding voltage

design, the current gain and the V_{CE0} breakdown voltage of the pnp transistor have to be extracted. The high-current characteristic of the diode string is shown in Figure 2.14 and a possible protection concept is illustrated in Figure 2.15.

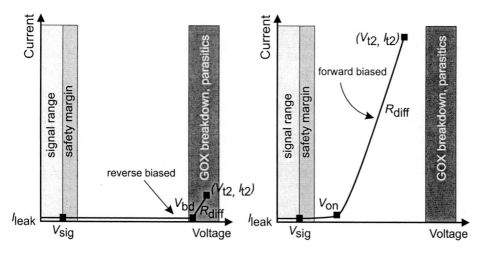

Figure 2.14: High-current characteristic of a diode string in forward and reverse-biased mode. The on-voltage V_{on} can be tuned by the number of diodes in series.

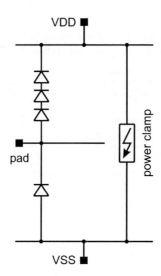

Figure 2.15: ESD protection concept using diode strings. As for single diodes, an appropriate power clamp is required to avoid stressing the diode in the reverse biased direction.

2.4.3   FET devices with active turn-on

Based on the NFET, further protection elements have been developed. By controlling the gate potential, the turn-on behaviour of the protection device can be optimized (Duvvury, 1995; Ker, 1996). These devices are called gate-coupled NFETs (gcNFETs). The gate control can be realized by an RC coupling, in which a high pass circuit allows the turn-on of the device when it is hit by the transient of the ESD pulse (Figure 2.16 a)). However, the application is limited to interfaces where the signal frequency range is significantly below the frequencies of the ESD pulse (Table 2.5). Often these elements are used as power clamps. Another version of gcNFET detects the over-voltage at the protected node, e.g. by the breakdown of a Zener diode, and actively triggers the device (Figure 2.16 b)). In comparison to the ggNFET, the high-current IV characteristic of the gcNFET is advantageous (see Figure 2.17). First, the trigger voltage V_{t1} is smaller than the V_{t1} of the NFET, resulting in less overvoltage stress before the device is turned on. Secondly, because V_{t1} is lower, I_{mf} is also reduced. As a consequence, homogeneous turn-on of multi-finger structures can be achieved with fewer layout measures and hence less area consumption. The parameters of actively controlled FETs can be found in Table 2.6.

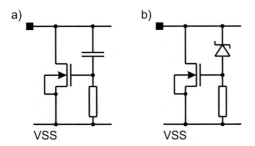

Figure 2.16: NFET with gate turned-on during an ESD event. a) *RC* coupling. b) Zener-biased circuit.

Table 2.5: Frequency range of the different ESD stress models (Amerasekera, 2002).

ESD model	center frequency	band width (6 dB)
HBM	500 kHz	2.1 MHz
MM	7 MHz	12 MHz
CDM	600 MHz	1.1 GHz

Table 2.6: Parameters of actively controlled FETs.

Parameter	Explanation
I_{leak}	leakage current at maximum operating voltage and temperature
V_{bd}	breakdown voltage at low current and low temperature
dV/dt	critical signal slope at maximum signal voltage and current
V_{t1}	turn-on voltage of parasitic bipolar transistor
I_{t1}	trigger current of parasitic bipolar transistor
V_h, I_h	holding voltage and holding current
R_{diff}	differential resistance in the low ohmic state of the device
I_{t2}, V_{t2}	current level that causes destruction of the protection element and corresponding voltage
V_{clamp}	(clamping) voltage at I_{ESD}

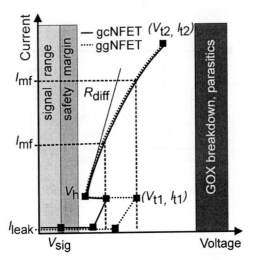

Figure 2.17: High-current characteristic of a gcNFET in comparison to a ggNFET (grey).
The main difference is the lower trigger voltage V_{t1} and, as a consequence,
the drastically reduced multi-finger trigger current.

2.4.4   Snapback protection devices

Besides FETs, there are other protection devices showing a snapback in the high
current IV characteristic. Prominent examples are:

- pnp or npn bipolar transistors as vertical and lateral devices (Jaffe, 1990;
  Johnson, 1993; Corsi, 1993; Gossner, 1999);

- bipolar transistors with an external triggering (Streibl, 2002; Amerasekera,
  1992);

- single SCRs (silicon controlled rectifiers, thyristors), see Figure 2.18 (Avery,
  1983; Ker, 1992; Duvvury, 1995);

- triggered SCRs, e.g., low-voltage triggered SCRs (LVTSCRs, triggering initiated by NFET transistors, see Figure 2.19), or SCRs triggered by Zener
  diodes (Chatterjee, 1991a,b; Russ, 2001).

Although these devices are based on totally different physical mechanisms, the
basic set of parameters of the characteristic in the high-current regime is essentially the same. To illustrate this, the high-current IV characteristic of an SCR is
depicted in Figure 2.20. Beside the ESD robustness itself, the trigger voltage and
current and the holding voltage and current are the crucial parameters (Table 2.7).
One condition for successful protection is that the ESD protection element trig-
gers before the protected element ($V_{t1,\mathrm{ESD}} < V_{t1,\mathrm{active}}$). Additionally, the holding
voltage of the protection element must not fall below the maximum signal voltage
($V_h > V_{\mathrm{signal,max}}$) or alternatively the holding current must be sufficiently high.
Otherwise, there is a risk of an unintentional latching during normal operation, as
a result, for example, of fast signal slopes or over- and undershoots. The risk of

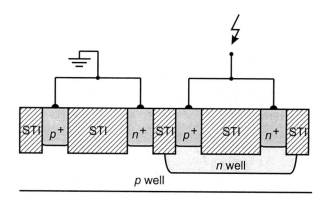

Figure 2.18: Cross section of a standard SCR in a technology using STI.

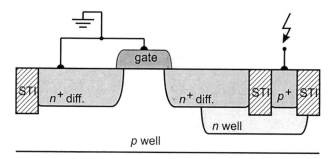

Figure 2.19: Cross section of a low-voltage triggered SCR (LVTSCR).

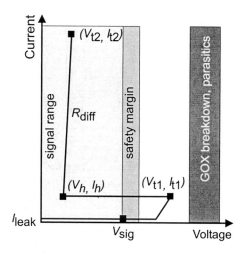

Figure 2.20: High-current IV characteristic of a simple SCR used as an ESD protection
element.   Without special design measures, the holding voltage is usually
in the signal range.   Special attention must therefore be paid to ensuring a
sufficiently high holding current, so as to prevent unintentional triggering
during normal operation.

Table 2.7: General parameters of snap-back devices in the high-current regime.

Parameter	Explanation
I_{leak}	leakage current at maximum operating voltage and temperature
V_{bd}	breakdown voltage at low current and low temperature
V_{t1}, I_{t1}	trigger current and turn-on voltage of inherent element
V_{h}, I_{h}	holding voltage and holding current
R_{diff}	differential resistance in the low ohmic state of the device
I_{t2}, V_{t2}	current level that causes destruction of the protection element and corresponding voltage
V_{clamp}	(clamping) voltage at I_{ESD}
$\mathrm{d}V/\mathrm{d}t$	triggering behaviour for fast transients if $V_{\text{sig}} < V_{\text{h}}$

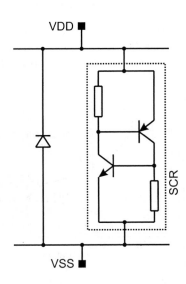

Figure 2.21: Protection concept using an SCR. The diode is mandatory to shunt the current negative/VDD.

latching is particularly high for "ideal" SCRs whose minimum V_h amounts only to a diode threshold plus the voltage drop across the charge-modulated region (Seitchik, 1987), i.e. ≈ 1.0–1.4 V. Special layout measures are necessary to shift the sustaining voltage or current to higher levels and prevent dV/dt triggering. The maximum current which might be injected into the relevant I/O cell must be limited to a value smaller than I_h. This problem vanishes for circuits operating at very low voltages (< 2 V). In general, devices such as SCRs have to be combined with a forward biased diode to protect their reverse operation (Figure 2.21).

2.4.5 Resistors

Resistors are used in ESD protection circuits to limit the current in ESD critical paths like the output drivers. There are different resistor implementations in use: a) well resistor (Carbajal, 1992), b) poly resistor (Banerjee, 1998), c) diffusion resistor (Krieger, 1989; Sanchez, 1999). For the last two types, the silicide formation has to be blocked to achieve a reasonable square resistance. With resistors a) and c) placed in the substrate, there is an increased risk of a breakdown to the substrate, if a high voltage appears at one head of the resistance. This occurs especially for high ohmic resistors.

The IV characteristic of the resistor is shown in Figure 2.22, the relevant parameters are listed in Table 2.8. A region of linear dependence between voltage and current density is observed at low current densities as typical ohmic behaviour, ideally described by Ohm's law

$$J = nq\mu_n V/L \qquad (2.2)$$

where n is the carrier density in the resistor, $q = 1.6 \times 10^{-19}$ C the electronic charge, μ_n the mobility of electrons (μ_p the mobility for holes in analogy for p-type resistors) and L the distance between the resistor contacts. Any contribution

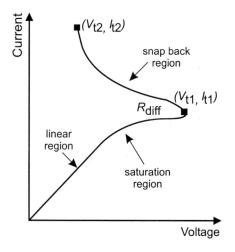

Figure 2.22: High-current IV characteristic of an n well resistor.

Table 2.8: General parameters of a resistor in the high-current regime.

Parameter	Explanation
R_{low}	resistance at low current and room temperature
R_{diff}	differential resistance below the triggering point
V_{bd}	break down voltage of the resistor head
$V_{\text{t1}}, I_{\text{t1}}$	trigger current and voltage
$I_{\text{t2}}, V_{\text{t2}}$	current level that causes destruction of the protection element and corresponding voltage
V_{clamp}	voltage drop across the resistor at I_{ESD}

from the resistor heads by inhomogeneous current flow or contact resistance is neglected. At high electric fields ($\approx 10^4$ V/cm in silicon) velocity saturation of the charge carriers is reached (e.g. 10^7 cm/s for electrons in silicon), which shows up as decrease in the slope of the IV characteristic. At even higher electric fields, avalanche multiplication starts and a snapback occurs as a result of the appearance of charge modulated filaments. If the filamentation is very pronounced, as in well resistors, the very high current density in the filament leads to local melting and damage to the resistor (Khurana, 1966; Hower, 1970; Yang, 1993). In low ohmic resistors, such as highly doped poly resistors, this condition (i.e. the saturation) might not be achieved, because the high current density leads to a thermal overload before saturation can occur (Amerasekera, 1993). The influence of thermal effects can be investigated by varying the pulse duration.

2.5   Requirements for a successful ESD protection concept

2.5.1   Requirements for ESD protection of output cells

In the simplest circuits, driver stages may be designed with only one PFET to VDDP (VDDP = VDDPower) and one NFET with source to VSSB (VSSB = bulk potential). The p well of the NFET is then connected to VSSB, whereas the n well of the PFET is tied to VDDP. Depending on the specification of the interface, the driver circuitry may be much more complex. Some possible requirements and the resulting changes to the circuitry are:

- Over-voltage tolerant pads (i.e. the signal voltage V_{sig} may exceed the power supply voltage VDDP by $V_{\text{sig}} - \text{VDDP} > 0.3$ V). Obviously, for this pad type there must be no p^+n diode between pad and VDDP (restriction for the ESD concept). Consequently, the n well of the PFET must not be connected directly to VDDP ("floating well concept").

- Signal voltages which exceed the reliability voltage limit of single devices. A common practice for overcoming the reliability issues is to use stacked

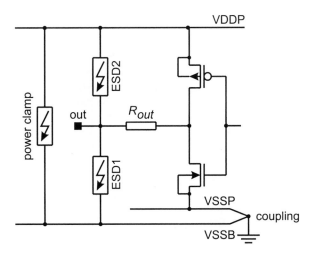

Figure 2.23: Typical output pad with substrate separation. The driver is connected to a "noisy" ground VSSP. In this example the ESD protection elements are tied to the bulk supply VSSB. Consequently, a low-ohmic coupling between VSSB and VSSP is mandatory for an elaborated ESD protection concept.

devices and control circuits for the gates of the devices connected to the pad.

- Interface specifications may cause overshoots/undershoots on the pad (such as reflection on the bus as defined in the PCI specification), causing temporarily over-voltages. Additional diodes may be required to shunt the over-swings.

- Substrate separation may be required in order to prevent switching noise coupling into the substrate which may disturb sensitive circuit blocks (e.g. PLL, ADC). For this reason, the source of the n driver is connected to a separate bus (commonly labelled as VSSP = VSSPower) which is not directly shortened on-chip with the VSSB bus.

An example for a schematic of a typical output pad with substrate separation is shown in Figure 2.23. The ESD protection elements ESD1 and ESD2 are in parallel to the driver transistors and the output resistor. In order to avoid a significant voltage drop at the output, the output resistor must be low-ohmic, or may even be absent.

The situation where because of performance reasons no resistor can be placed between the pad and the driver will be discussed in the following, because this is the most critical situation for the ESD designer. Between pad and VSSB/VDDP there might be additional parasitic elements such as decoupling capacities. The desired IV characteristic of the ESD protection element and the driver is shown schematically in Figure 2.24. In general, the characteristic of n and p drivers and of any protection element is defined by a small set of parameters in the breakdown and high-current regime. Of course, the values of the parameters depend on the

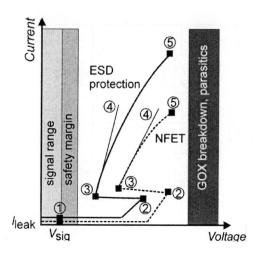

Figure 2.24: High-current IV characteristic of parallel protection elements and an active
device such as an NFET. This example represents the typical situation in an
output pad without a resistor between pad and driver. From this schematic,
the rudimentary ESD guidelines for an output pad can easily be deduced.
The decisive set of parameters are (1) I_{leak}, (2) V_{t1}, I_{t1}, (3) V_{h}, (4) R_{diff}, (5)
I_{t2}, V_{t2} for both ESD protection element and active device.

type of the element; for example, an n driver has usually a much stronger "snapback" $(V_{\text{t1}} - V_{\text{h}})$ than the corresponding p driver in the same technology. Using
Figure 2.23 and 2.24, the requirements for the ESD protection concept and the
ESD protection elements can be deduced from very basic considerations (see also
for example Amerasekera 2002; Dabral 1998; Esmark 2000).

1. At any time during the stress, the current forced through the driver must
   not exceed the acceptable maximum current. Either the additional ESD
   protection element must shunt the main portion of the ESD current, or the
   active element must be robust enough.

   (a) If a self-protecting concept is intended, the device width/device layout
       must be chosen to guarantee a sufficiently high I_{t2}. In most contempo-
       rary sub-µm technologies, the self-protecting concept can only be used
       if the layout of the driver is ESD-optimized (for example by using a
       ballasting resistor at drain/source).

   (b) If an ESD protection element is used in parallel to the driver, the ratio
       of the current through the ESD protection and the active element can
       be deduced from Figure 2.24 for a certain total ESD current. This ratio
       can be tuned by varying the type, layout and width of the protection
       element and possibly by a serial resistor R_{out}.

2. The ESD protection element itself must not be damaged by the ESD current
   $(I_{\text{t2,protection}} > I_{\text{ESD}})$. A safety margin accounting for the variation of the
   process technology has to be designed in.

3. The ESD protection element must trigger before the active element turns on ($V_{t1,\text{protection}} < V_{t1,\text{active}}$). Transient turn-on of the active element must be prevented, otherwise measures to increase the ESD robustness of the active device itself must be implemented.

4. If the trigger current is exceeded, the voltage of the protection element in the high-current regime (and the drivers) should not fall below the maximum signal voltage ($V_h > V_{\text{signal,max}}$). Otherwise, there would be a risk of an unintentional switch-on during operation owing to signal overshoots or fast transients.

5. The maximum voltage during ESD stress (V_{clamp} or V_{t1}), where V_{clamp} is given by

$$V_{\text{clamp}} = V_h + \int_{I_h}^{I_{\text{ESD}}} R_{\text{diff}}(I) \, \mathrm{d}I \qquad (2.3)$$

must not exceed the minimum breakdown voltage of the parasitic elements. The voltage clamping capability V_{clamp} can be tuned, for example by varying the device width of the protection element or by using an ESD protection element with a lower holding voltage.

6. During operation, the leakage current of all elements of the output cell (ESD protection elements and drivers) must not exceed a certain limit defined in the output cell specification. Additional requirements might be the maximum area consumption and the maximum capacitive load (in the case of high-frequency applications) of the output cell elements.

7. If a substrate separation is provided, there must be a low-ohmic coupling between VSSB and VSSP to establish a closed protection path between the output and any ground pin. Often a double bond acts as an inductive decoupling for high-frequency noise, can be used to shorten VSSB and VSSP. The maximum bus resistance between the signal pad and the VSSB/VSSP pin must be limited in order to avoid an excessive additional voltage drop across the bus which (a) might lead to a parasitic breakdown and (b) forces more current through the active devices.

2.5.2 Requirements for ESD protection of input circuits

A typical digital input cell is shown in Figure 2.25. The important difference from the output cell is that gates are connected to the pad instead of diffusions. Gates can only carry a very limited current in the breakdown (typically 6–12 orders of magnitude smaller than the ESD stress current) and are thus easily damaged, if over-voltage appears at the gate node during ESD. Serial resistors between gate and pad of several hundred Ohms are usually tolerable in the input path. This allows implementation of a two-stage protection concept for better voltage clamping in the input circuits. Basically, the primary protection stage is identical to the output cells. In practice, the clamping requirement for the primary stage is even more relaxed. The second stage provides effective voltage clamping at the

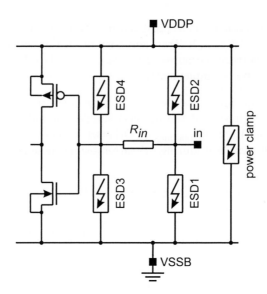

Figure 2.25: Typical input pad of an integrated circuit containing a two-stage ESD protection concept. The second stage with protection elements ESD3 and ESD4 are de-coupled from the pad by R_{in} and provide a safe voltage clamping for the gate node.

gate node and prevents a breakdown of the gate oxide or other parasitic elements connected to that node. The serial resistor R_{in} limits the current through the second stage. Usually, either small NFET devices or small diodes act as voltage clamps for the gates. In addition to requirements identical to points (2), (4), (5), and (6) of the output pads (see above), the following considerations must be taken into account for input pads.

8. Although most of the ESD current is shunted by the primary ESD protection elements, the resistor R_{in} and the elements of the second stage must be capable of withstanding a certain current. The current through R_{in} can be estimated from the *IV* characteristics in the high-current regime (see Figure 2.24) by

$$I_{\mathrm{ESD},2} = \frac{V_{\mathrm{h},1} - V_{\mathrm{h},2} + I_{\mathrm{ESD}} R_{\mathrm{diff},1}}{R_{\mathrm{diff},1} + R_{\mathrm{in}} + R_{\mathrm{diff},2}} \qquad (2.4)$$

where index "1" denotes a parameter of the primary ESD protection element, index "2" a parameter of the second stage. Using the same protection element for the primary ESD protection and the second stage ($V_{\mathrm{h},1} = V_{\mathrm{h},2}$) and assuming $R_{\mathrm{in}} \gg R_{\mathrm{diff},1}$, $R_{\mathrm{in}} \gg R_{\mathrm{diff},2}$, $I_{\mathrm{ESD},2}$ is given by

$$I_{\mathrm{ESD},2} = \frac{I_{\mathrm{ESD}} R_{\mathrm{diff},1}}{R_{\mathrm{in}}} \qquad (2.5)$$

The ESD withstand current of R_{in} and the second stage must be adjusted by the width of the devices to be compliant with $I_{\mathrm{ESD},2}$.

9. Both the clamping voltage V_{clamp} and the trigger voltage V_{t1} of the second stage must be sufficiently low to prevent breakdown of the gate oxides of the inverter stage and further possible parasitics. The voltage clamping capability can be optimized by increasing R_{in}, and/or increasing the device width of ESD 3/ESD 4, and/or choosing a second stage with optimized breakdown/snapback behaviour.

2.5.3   Requirements for ESD protection of the core

ESD stress applied to an I/O cell may not only affect the devices connected directly to the pad, but also the core. This is even more of a risk if the ESD event directly includes the supply pads. ESD problems in the core are mostly caused by a high potential drop between VDD and VSS or different supply domains VDDx, VDDy. Two potential risks are known:

10. Breakdown between parasitics. Such parasitics may be, for example (see Duvvury 1988; Johnson 1993):

    - neighbouring n^+/n^+ diffusions (Figure 2.26), n^+ diffusions close to n well, neighbouring n well/n wells connected to different supply domains,
    - n^+ diffusions/p^+ diffusions,
    - gate oxides, such as those used in capacitors or other thin dielectrics between different supplies,
    - breakdown between in-plane metal Mx/metal Mx and breakdown between different metal layers metal Mx/metal My.

    In order to avoid breakdown, the maximum voltage drop between VDD and VSS (or VDDy) must not exceed the minimum breakdown voltage for any possible combination of parasitics. The voltage drop between VDD and VSS (VDDy) can be calculated from the high-current characteristic of the power supply clamp, the maximum allowed bus resistance, and the high-current behaviour of the coupling elements between power supply domains.

11. Damage of NFETs in inverter structures with large PFETs. This problem has gained significant importance in deep-sub-μm technologies. The failure mechanism can be explained as follows (see Figure 2.27). During an ESD stress, the potential on VDD can be significantly higher than during normal operation. The maximum voltage drop across the PFET can be estimated by $V_{\text{PFET}} = \text{VDD}_{\text{ESD}} - V_{\text{h,NFET}}$, where $V_{\text{h,NFET}}$ is the holding voltage of the NFET and gives the lowest possible voltage across the NFET in the breakdown region if a current larger than the trigger current is flowing. Depending on the gate potential of the PFET, the PFET can drive a very high current into the NFET, which is proportional to $W_{\text{PFET}}/L_{\text{PFET}}$. In the worst case only the smallest finger of the NFET triggers and limits the ESD performance of the entire NFET. In order to prevent such fails, $\text{VDD}_{\text{clamp}}$ must be adjusted to a sufficiently low value. However, this measure costs more effort and area for the power clamp element and the power bussing.

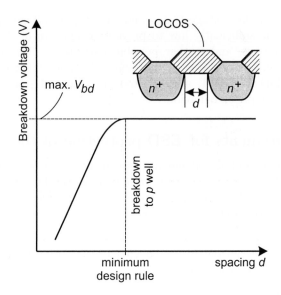

Figure 2.26: Breakdown voltage in dependence on the spacing d between two n^+ diffusions forming a parasitic bipolar. Below a minimum spacing d defined in the design guidelines, the breakdown voltage is decreased. Consequently, the ESD design window is decreased, too.

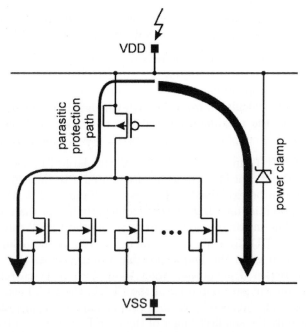

Figure 2.27: Inverter-type structure which is endangered during an ESD event in case of non-sufficient power clamping. Because of the non-ESD optimized layout of core devices, the current might be shunt through one single of the NFET.

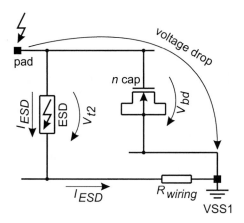

Figure 2.28: Voltage drop over a parasitic device (here an NFET is used as a capacity) during an ESD event due to parasitic resistances in the bus system.

2.5.4 Wiring

Metal wires, vias and contacts, not to mention the power supply concepts and the coupling between the different power supply domains, may all have an influence on the ESD behaviour of products. Three issues have to be considered.

12. In any application, the path of an ESD stress can be represented by pad → metal wires → vias/contacts → ESD protection element/driver → vias/-contacts → metal buses → pad. Obviously, the metal connections, vias, and contacts must be capable of withstanding the ESD stress without any degradation.

13. The resistance of the wiring also plays an important role during ESD. Because of the ESD current, which may be 2–3 orders of magnitude higher than the currents flowing in the operational regime, wiring resistances cause a significant additional voltage drop (see, for example, Figure 2.28). The maximum allowable resistance of all connections can be estimated by

$$R_{\text{wiring}} < \frac{V_{\text{bd,parasitics}} - V_{\text{clamp}}}{I_{\text{ESD}}} \qquad (2.6)$$

where R_{wiring} denotes the sum of all metal/via/contact resistances. In a proper ESD design the resistance of vias and contacts can be neglected because already the current capability requires a large amount of parallel contacts and vias; the resistance is then determined by the metal bus resistance between the ESD protection element and the power supply pad.

14. Because the ESD stress may occur between pads connected to different power supplies, a sufficient coupling between the power supply domains is a crucial issue. As the various VDDx often cannot be coupled via anti-parallel diodes or metal connections due to their different potentials, it is essential to have

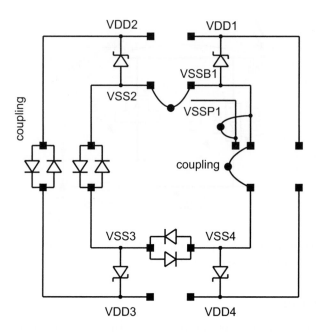

Figure 2.29: ESD protection scheme of an integrated circuit with several power supply
domains.

very good coupling of the ground rail. A typical example for a power supply
coupling is shown in Figure 2.29. Power supply block "1" has substrate
separation, commonly used in digital pads with relatively large and fast
drivers. Power supply domains "2", "3", and "4" represent something like
analog blocks or core supply. Between block "2" and "3" coupling of VDD
is possible. All VSSx are coupled to the neighbouring block, either by a
double bond (metallic connection, e.g. between "1" and "2" and between
"1" and "4") or by a pair of anti-parallel diodes (between "2" and "3" and
between '3' and '4'). In each power supply domain there must be sufficient
voltage clamping during an ESD stress. The voltage clamping is provided
by a power supply clamp. Alongside a high ESD withstand current, the
voltage clamping capability is its crucial parameter. Examples for power
supply clamps are grounded-gate NFETs, RC-triggered NFETs, SCRs and
diode strings.

Summary

- Typical current densities in devices driven into breakdown during ESD are
  10–100 mA/μm^2. All these elements must be evaluated for a protection
  development in this high current regime.

- The protection capability of a protection device under ESD stress is described by the high-current IV characteristic including the corner values of

breakdown voltage, trigger voltage, holding voltage after snapback, clamping voltage at the ESD current and failure threshold. This high-current characteristic must fit in the ESD design window. The lower limit of the ESD design window is defined by the maximum ratings of the signal voltage and the upper limit by the minimum critical breakdown voltage of protected devices, e.g. the gate oxide breakdown voltage.

- The conditions for good ESD robustness can be fulfilled both by the active circuitry itself, in the so-called self-protection concept, or by adding additional shunt elements showing IV characteristics compliant with the ESD design window.

- In making the right choice of protection element among the many possibilities, such as diodes, grounded gate NFETs, bipolar transistors etc., it is important to consider their parameters and compare them with the circuit requirements.

- A complete ESD protection concept for an IC must consider factors beyond the local protection; for example, for an I/O cell, the coupling between the power supply domains must also be considered, so as to provide a safe discharge path between any combination of pins.

Bibliography

Amerasekera A., Chatterjee A., "An Investigation of BiCMOS ESD Protection Circuit Elements and Applications in Submicron Technologies", Proc. 14th EOS/ESD Symposium (1992), 265.

Amerasekera A., Chang M.-C., Seitchik J., Chatterjee A., Mayaram K., Chern J.-H., "Self-heating Effects in Basic Semiconductor Structures", IEEE Electr. Dev. (1993), 1836.

Amerasekera A., Duvvury C., *ESD in Silicon Integrated Circuits, Second Edition,* John Wiley, Chichester, England, 2002.

Avery L.,"Using SCRs as Transient Protection Structures in Integrated Circuits", Proc. 5th EOS/ESD Symposium (1983), 177.

Banerjee K., Amerasekera A., Kittl J., Hu C., "High Current Effects in Silicide Films for Sub-0.25 μm VLSI Technologies", Proc. IRPS (1998), 284.

Carbajal B., Cline R., Andersen B., "A Successful HBM ESD Protection Circuit for Micron and Sub-Micron Level CMOS", Proc. 14th EOS/ESD Symposium (1992), 234.

Chatterjee A., Seitchik J.A., Chern J.-H., Wang P., Wei C.-C, "Direct Evidence Supporting the Premises of a Twodimensional Diode Model for the Parasitic Thyristor in CMOS Circuits Built on Thin Epi", IEEE Trans. Electr. Dev. (1988), 509.

Chatterjee A., Polgreen T., "A Low-Voltage Triggering SCR for On-chip Protection at Output and Input Pads", Electr. Dev. Lett. (1991a), 21.

Chatterjee A, Polgreen T., Amerasekera A., "Design and Simulation of a 4 kV ESD Protection Circuit for a 0.8 μm BiCMOS process", Tech. Dig. IEDM (1991b), 913.

Corsi M., Nimmo R., Fattori F., "ESD Protection of BiCMOS Integrated Circuits which Need to Operate in the Harsh Environments of Automotive or Industrial", Proc. 15th EOS/ESD Symposium (1993), 209.

Dabral S., Maloney T.J., *Basic ESD and I/O Design* John Wiley, New York, 1998.

Duvvury C., Rountree R.N., Adams G.,"Internal Chip ESD Phenomena beyond the Protection Circuit", IEEE Trans. Electr. Dev. (1988), 2133.

Duvvury C., Amerasekera A., "Advanced CMOS Protection Device Triggermechanisms During CDM", Proc. 17th EOS/ESD Symposium (1995), 162.

Esmark K., Stadler W., Gossner H., Wendel M., Guggenmos X., Fichtner W., "Advanced 2D/3D ESD Device Simulation – a Powerful Tool Already Used in a Pre-Si Phase", Proc. 22nd EOS/ESD Symposium (2000), 420; Microelectronic Reliability **41** (2001), 1761.

Gossner H., Müller-Lynch T., Esmark K., Stecher M., "Wide Range Control of the Sustaining Voltage of ESD Protection Elements Realized in a Smart Power Technology", Proc. 21st EOS/ESD Symposium (1999), 19.

Herlet A., Raithel K., "Forward Characteristics of Thyristors in the Fired State", J. Solid State Electr. (1966), 1089.

Hower P.L., Gopala K.R.V., "Avalanche Injection and Second Breakdown in Transistors", IEEE Trans. Electr. Devices (1970), 320.

Jaffe M., Cottrell P.E.,"Electrostatic Discharge Protection in a 4-Mbit DRAM", Proc. 12th EOS/ESD Symposium (1990), 218.

Johnson C.C., Qawami S., Maloney T.J.,"Two Unsusual Failure Mechnisms on a Mature CMOS Process", Proc. 15th EOS/ESD Symposium (1993), 225.

Ker M.D., Wu C.Y., Lee C., "A Novel CMOS ESD/EOS Protection Circuit with Full-SCR Structures", Proc. 14th EOS/ESD Symposium (1992), 258.

Ker M.D., Wu C.Y., Cheng T., Chang H.H., "Capacitor-Coupled ESD Protection Circuit for Deep-Submicron Low-Voltage CMOS ASIC", IEEE Trans. VLSI Systems (1996), 307.

Khurana B.S., Sugano T., Yanai H., "Thermal Breakdown in Silicon p-n Junction Devices", IEEE Trans. Electr. Devices (1966), 763.

Krieger G., "The Dynamics of Electrostatic Discharge Prior to Bipolar Action Related Snapback", Proc. 11th EOS/ESD Symposium (1989), 136.

Maloney T.J., Dabral S., "Novel Clamp Circuits for IC Power Supply Protection", Proc. 18th EOS/ESD Symposium (1995), 1.

Russ C.C., Mergens M.P.J., Armer J., Jozwiak P.C., Kolluri G., Avery L.R., Verhaege K.G., "GGSCRs: GGNMOS Triggered Silicon Controlled Rectifiers for ESD Protection in Deep Sub-Micron CMOS Processes", Proc. 23rd EOS/ESD Symposium (2001), 22.

Sanchez H., Siegel J., Nicoletta C., Alvarez J., Nissen J.,"A Versatile 3.3 V/2.5 V/1.8 V CMOS I/O Driver Built in a 0.2 µm 3.5 nm Tox 1.8 V Technology", Tech. Dig. ISSCC (1999), 276.

Seitchik J., Chatterjee A., Yang P., "An Analytical Model of Holding Volatge for Latchup in Epitaxial CMOS", IEEE Electr. Dev. Lett. (1987), 157.

Streibl M., Esmark K., Sieck A., Stadler W., Wendel M., Szatkowski J., Gossner H., "Harnessing the Base-Pushout Effect for ESD Protection in Bipolar and BiCMOS Technologies", Proc. 24th EOS/ESD Symposium (2002), 73.

Voldman S.H.,"ESD Protection in a Mixed Voltage Interface and Multi-Rail Disconnected Power Grid Enviroment in 0.5- and 0.25-µm Channel Length CMOS Technologies ", Proc. EOS/ESD Symposium (1994), 125.

Voldman S.H., Gerosa G., Gross V.P., Dickson N., Furkay S., Slinkman J. "Analysis of Snubber-Clamped Diode-String Mixed Voltage Interface ESD Protection Network for Advanced Microprocessors", Proc. 17th EOS/ESD Symposium (1995), 43.

Yang P., Chern J.-H., "Design for Reliabilty: The Major Challenge for VLSI", Proc. IEEE (1993), 730.

Chapter 3

Simulation flow

3.1   ESD protection development strategy

Any ESD protection development is constrained by the performance requirements of the protected circuits and the limits of the process technology. In the development of a technology, the focus is mainly on the optimization of the active devices like FETs. ESD-relevant parameters and ESD protection devices often only receive minor attention. In many cases, high intrinsic ESD robustness of the transistors is considered as an optional feature, not mandatory for the process release.

While the development of a technology is still in the status of optimization, the pilot product already requires the delivery of an ESD concept and I/O cells. The ESD development strategy has to take this into account. Therefore, an early assessment of the technology's properties and their correlation with the ESD protection capability of the chosen concept is a key factor in minimizing later changes.

The question might arise whether it is more reasonable to wait for a mature technology phase before the ESD concept is implemented and I/O cells are delivered. However, there are good reasons why concurrent engineering of the circuits and the technology is required.

At present, sub-0.1 μm CMOS logic technologies are in development. Products with a complexity of several hundred million gates per integrated circuit will be realized in these technology generations. The cost per mask set for a reticle will be far beyond $ 1 million. Thus, any test run will be extremely expensive, especially when 12" substrates are used. The demands of both complexity and cost require as early a start as possible to the pilot products and the sharing of reticles among technology and product development, in order to keep mask and processing costs as low as possible.

In addition, the short lifetime cycles of modern mass products like those in the wireless communication market, driven by stringent cost and innovation considerations, also require an early start to product development. Often a product is fabricated for only 1–2 years in one technology, at the end of which period cost reduction and/or innovation pressure force a move to the next technology generation.

Now proceeding on the technology roadmap in conjunction with the approach of concurrent engineering, ESD development faces several severe issues.

- The aggressive time and cost frame for the development of new products require a "first time right" for most of the functionality and features of the IC, including the ESD concept. Non-working ESD concepts, in particular where they affect the testability of the IC functions, (e.g. by leakage paths in the I/O cells), necessitate a full mask redesign and the loss of a complete learning cycle.

- Scaling down the technology produces smaller and faster circuits. This puts a lot of pressure on the ESD engineer, who has to deal with non-scalable ESD effects. The percentage of area used in the I/O cells and the degradation of performance due to ESD protection measures become increasingly important issues.

Unfortunately there is no universal solution for ESD protection which fits all technologies and circuit conditions. Each concept has its advantages and disadvantages, see Chapter 2.

Self protection by larger NFET and PFET devices allows straightforward realization without additional effort for the monitoring and design environments (Gauthier, 2001), but might consume significant additional area for I/O cells with smaller drivers. Specific silicon controlled rectifiers have a high area efficiency, but often require special treatment in the design kit and careful monitoring of the technology (Amerasekera, 2002). Diodes used under forward biased condition are robust and small, but are hardly applicable for overvoltage tolerant circuits and have to be combined with efficient power clamp elements (Maloney, 1995; Gauthier, 2001). Lateral bipolar transistors work well for certain LOCOS technologies, but fail dramatically in shallow trench isolation (STI) processes (Never, 1995). A vertical *pnp* transistor might be an attractive ESD protection element in technologies featuring an epi substrate, but it is generally not applicable for non-epi technologies (Amerasekera, 2002).

Stepping from one technology to the next is a major challenge for ESD development, as in general there can be no simple and smooth transfer of protection concepts from the preceding technology. In the past, ESD concepts suddenly failed as a result of the introduction of new process features such as the lightly doped drain (LDD) or the STI (Amerasekera, 2002). In the future, there are numerous examples of tightened process constraints that the ESD engineer will have to deal with. The reduction of the gate oxide thickness and the introduction of new gate dielectrics, leading to low breakdown voltages, the increased well resistance, preventing the construction of reasonably low ohmic diodes, or the snapback of PFET devices with very short gate length, might all cause surprises, not to mention the totally different situation with fully depleted silicon on insulator (FD-SOI) technologies.

To overcome all these obstacles and stumbling stones, modern ESD development employs a broad spectrum of simulation and analysis methods. The ESD development flow, as understood here, describes the sequence and the links between the single methods and work steps, and the interaction between ESD engineer and technology development and circuit design. Typically the phases of any ESD development flow can be divided into three major parts:

1. pre-silicon concept definition and test structure development;

2. test structure analysis and deduction of an optimized concept;

3. verification of the concept at I/O cell or product level.

One essential measure of the quality of the development flow is whether or not the pre-silicon concept is close to the final design. On the one hand, the pre-silicon concept has to guarantee full operational functionality and a reasonable ESD performance for the engineering samples, while the technology is still under construction, and this requires a more robust design with larger margins. On the other hand, overly large safety margins in the final optimized concept causes area loss for products in high volume production. However, a complete rework of the I/O cells and other ESD relevant circuit parts for the area and performance optimized version costs a lot of effort and delays the final verification of the complete I/O cells.

The phases of a common development flow break down into the following work steps (see the upper part of Figure 3.1):

- Use the know-how from previous technologies to collect the ESD relevant parameters of the new technology. The parameters cover not only straightforward electrical parameters like breakdown voltage and sheet resistance, but also detailed process information like the LDD profile. The list of parameters depends on the protection philosophy and is very technology-line-specific. It is certainly different for CMOS and bipolar technologies and might even vary between different fabrication sites. The most important parameters for typical protection concepts are discussed in Chapter 2.

- Collect the circuit requirements, for example signal voltage range, leakage constraints and lifetime.

- Assess possible protection concepts and decide about the pre-silicon concept. As some parameters are usually not yet available or very inaccurate in this early technology development stage, this will be a very intuitive task, making assumptions based on experience with previous technologies. Moreover, many precaution measures will be built in to reduce possible risks. This results in a non-optimum area consumption for the ESD protection measures.

- Design test structures including possible protection elements, protected circuit parts (like driver stages), critical parasitics (junctions, gate oxide, etc.) and metal reliability structures. This often leads to a very large number of test devices, as many protection devices in multiple layout variations have to be examined (SEMATECH, 1998).

- Analyse silicon and decide about the area and performance optimized concept. In the early phase of modern CMOS technologies the processing time,

time scale

empirical ESD development flow:

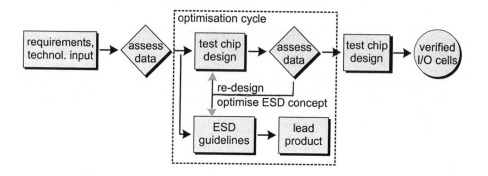

simulation based ESD development flow:

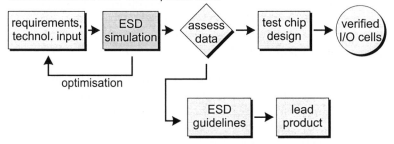

Figure 3.1: Representations of typical ESD development flows. Top: Standard flow based
on test chip analysis. Bottom: Advanced, simulation-based flow.

including mask generation, often takes more than half a year. This causes
a significant delay in the implementation of the area and performance optimized ESD concept. Owing to the large number of test structures, the
analysis time and effort is enormous. Several test cycles might be necessary
to get a satisfactory result.

- Give feedback to process technology development about how to guarantee
  and improve ESD performance by technology measures. Apart from some
  specific process steps like modification of the LDD, this is a difficult task since
  many of the 500 or more process steps in a modern CMOS process can have
  influence on the ESD performance. For cost and time reasons, only a very
  limited number of process split lots can be run for this purpose. In addition,
  a late change request in a technology might interfere with previously defined
  process features and, therefore, cannot be considered.

- Implement the area and performance optimized ESD rules in the I/O cells
  and products and verify the ESD hardness of these circuit systems.

The most critical issue in this approach is the limited knowledge of the ESD relevant process features at the beginning of the development. This often causes major reworkings before the final concept is achieved. This deficiency is the starting point for a simulation based ESD development flow as outlined in the lower part of Figure 3.1.

In contrast to the "empirical" flow, the simulation based development flow uses quantitative process data at an early stage. Based on the detailed description of the process steps, commercial process simulators can generate a 2D or even a 3D structure including all the information about materials, topology and doping profiles. Using the simulated structures as input for a device simulation, all essential ESD parameters can be examined. Usually the ESD design window is first determined by an analysis of PFET and NFET drivers. Based on this, the possible protection structures can be evaluated and optimized. The layout parameters are adjusted to meet the requirements of the ESD design window. Finally, proposals for the optimization potential of the technology itself can be extracted. The test structure kit can then be focused on a limited number of structures for design centring and calibration. Feeding the extracted electrical parameters into a circuit simulation under ESD conditions allows an evaluation of the quality of the complete protection circuit. Thus, the final ESD performance of the circuit and the technology parameter can be linked, allowing optimization of the ESD protection very early in the development.

3.2   Simulation based development flow

As discussed in the previous section, the correct design of an ESD protection concept is based on a large number of properly adjusted parameters. However, these parameters are often not within the scope of the device development in a new technology. Yet the values of these parameters are very sensitive to the exact process. The acceptable margins for the parameters are set by the circuit application.

Thus, the task of a successful ESD simulation is to evaluate the ESD parameters on the basis of the early process information and to check whether or not the parameter values comply with the request of the circuit. This demands a simulation flow (Figure 3.2) which starts with the process simulation creating the devices under investigation, leads to a device simulation providing the electrical parameters, and is followed by a circuit simulation proving the applicability of the protection devices in the circuitry and optimizing the protection circuit. Finally, the complete IC simulation verifies the product specific implementation of the ESD protection including the power supply concept. In this flow the output of the process simulation serves as input for the device simulation. The extracted parameters from the device simulation are the basis for a compact simulation of analogue circuits or I/O cells. This again provides IV characteristics for the behaviour of the complete cell which can be used for a chip-level ESD verification. In the following the status and the inputs and outputs of the various ESD simulation steps are highlighted.

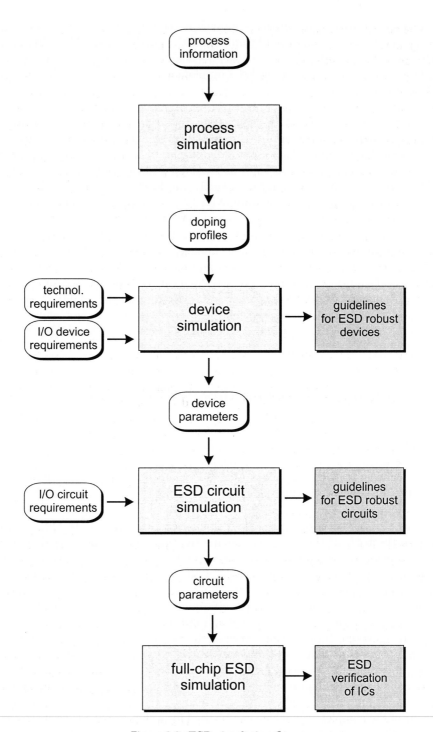

Figure 3.2: ESD simulation flow.

The process simulation is a well-established tool in modern semiconductor technology development. For most of the common process steps like implantation, diffusion, oxide growth, etc., evaluated models are available, see Chapter 4. However, in every new process technology the coefficients have to be re-calibrated. ESD-specific process steps, like I/O LDD implantation or silicide blocking, often need additional attention as they are not automatically included. The calibration of the models is based on electrical or physical measurements. Secondary ion mass spectroscopy (SIMS) provides information about vertical profiles, while SEM or transmission electron microscopy (TEM) of a cross section through the devices delivers the necessary data for the topology and layer thickness. By electrical measurements of test structures the lateral profile is probed. Based on the description of the process of record (POR) and the definition of the lithographic masks, the calibrated simulator produces a 1D, 2D or 3D structure including all the topology and material information mapped to an appropriate mesh for the numerical treatment. To investigate the robustness of the extracted ESD parameters concerning process variations, structures should be also created that consider the process window data. These files can be used as direct input into the device simulation. The quality of the profiles is most important for the reliability of the electrical parameters subsequently extracted.

For a 1D, 2D or 3D structure resulting from a process simulation, the device simulation calculates the electrical, thermal and optical response to certain stimuli, which are typically applied as voltage or current force to the predefined electrodes, see Chapter 5. There are several applicable commercial tools on the market, which differ in the set of implemented models and in the numerical solution approach, e.g. ATLAS™ (Silvaco International, Santa Clara, CA), DESSIS™ (ISE AG, Zurich, Switzerland), MEDICI™ (Synopsys Inc., Mountain View, CA). All the tools also allow investigation of more complicated topologies. In standard use, for CMOS technologies above 100 nm gate length, operating voltages of a few volts and temperatures of $< 400\,°C$ in the silicon, the tools are well proven and reproduce the experimental results, provided the calibration is done and the doping profiles are correct. At much higher temperatures or at very high current densities and voltages, many of the models are not properly verified. These limitations are discussed in Section 5.1. The calibration procedure involves comparison with the *IV* characteristics of test structures, both at low voltages and at voltages beyond the breakdown of the *pn* junctions. In addition to the calibration of the models, the numerical mesh also has to be adjusted to the parameters under investigation. For example, the density of the grid should be chosen reasonably high with respect to the occurring gradients of the electric field, impact ionization or lattice temperature inside the device under ESD stress. Finally, a lumped element network can be defined to simulate stimuli with specific wave forms like those appearing during an ESD discharge. The major output of the device simulation is the *IV* characteristic, but it also reveals the "invisible" internal parameters like current density distribution, temperature distribution, carrier density etc., which cannot be measured at the terminals. Depending on the problem, 1D, 2D or 3D distributions can be examined. The recently implemented 3D calculations, however, are very CPU time consuming.

Based on the (calculated) IV characteristics, parameters for lumped element models can be extracted either manually or by electrical analysis tools, e.g. IC-CAP (Agilent, Palo Alto, CA). The parameters are used for the circuit level simulation, which is a tool to evaluate a larger number of elements (typically for ESD simulation in the order of several dozens), see Chapter 6. There is a great number of various commercial and in-house tools available, for example, HSPICE®, SABER® (Synopsys Inc., Mountain View, CA), SPECTRE® (Cadence, San Jose, CA), TITAN (Infineon Technologies, Munich). However, the commonly used models for the simulation of FETs (for an overview see Massobrio 1993) do not include the ESD relevant behaviour. Neither bipolar action nor thermal effects are modelled. For this purpose ESD specific models have been developed and are permanently improved, for example, ET-SIM (Urbana), SQ3 (Bosch, Germany/Fraunhofer, Munich/Infineon). The models are calibrated using the IV characteristics of test structures with certain design variations. The circuit simulation allows evaluation of the current distribution between the protection path and the protected circuitry during the ESD event. Dynamic switching behaviour arising from fast transients can also be revealed. By determining the behaviour of a complex circuit under ESD conditions it is possible to generate a macro model of this circuit, which can be used in higher level simulations. As the lumped element models are based on an empirical extraction from a certain set of devices, their use for the investigation of devices concerning process variations, or design changes well beyond the measured range, is not recommended. For this purpose, device simulation is definitely the better tool with a more reliable physical basis.

A chip-level simulation including all the devices, the parasitics and the influence of the package cannot work as a straightforward application of a circuit simulation, because of the numerical problem with the extremely high number of elements, see Chapter 7. Based on the schematics containing ESD specific macro models and the physical layout, a simplified circuit has to be extracted which can be treated by a circuit simulation. However, a general methodology of extracting the simplified circuits without obliterating the effect under investigation has not yet been established. Finally the outputs from chip-level simulations are potential weak points in the design, where ESD failures might occur. This can be considered as the "ultimate" ESD verification of an IC before processing.

Summary

- The definition of a successful ESD protection concept subject to technology constraints and circuit requirements requires a well-thought out development flow, particularly as the ESD concepts have to be implemented while the process technology is in an early phase and is thus liable to change.

- Whereas, in a test structure-based development flow, the influence of technology changes can only be evaluated with the added delay of processing and measuring new samples, a simulation based flow can give more or less immediate feedback and provides proposals for optimized process designs without large test structure effort. This also offers the opportunity to obtain agreement between process constraints and required ESD relevant technology parameters as early as the process definition phase.

- An ESD simulation flow contains the process simulation of the devices under investigation, their electrical operation under ESD conditions by a device simulation, the extraction of compact models for the calculation of complete circuits like I/O cells and the chip-level simulation for a final verification of the ESD concept of the IC.

Bibliography

Amerasekera A., Duvvury C., *ESD in Silicon Integrated Circuits, Second Edition,* John Wiley, Chichester, England, 2002.

Gauthier R., Stadler W., Esmark K., Riess P., Salman A., Muhammad M., Putnam C., "Evaluation of Diode-Based and nMOS/Lnpn-Based ESD Protection Strategies in a Triple Gate Oxide Thickness 0.13 μm CMOS Logic Technology", Proc. 23rd EOS/ESD Symposium (2001), 205.

Maloney T., "Novel Clamp Circuits for IC Power Supply Protection", Proc. 17th EOS/ESD Symposium (1995), 1.

Massobrio G., Antognetti P., *Semiconductor Device Modeling with SPICE,* McGraw-Hill, New York, 1993.

Never J., Voldman S., "Failure Analysis of Shallow Trench Isolation ESD Structures", Proc. 17th EOS/ESD Symposium (1995), 273.

SEMATECH ESD Technology Working Group, "Test Structures for Benchmarking the Electrostatic Discharge (ESD) Robustness of CMOS Technologies", SEMATECH Technology Transfer # 98013452A–TR (1998).

Chapter 4

Process simulation

4.1 Introduction

The starting point for ESD device simulation is to construct an electronic pattern of the device as processed in the semiconductor technology. Such a pattern describes the topology of the materials used for the specific device. Most important materials in today's CMOS process technology are silicon (for the bulk of the semiconductor device), polysilicon (for the gates of NFETs and PFETs) and dielectrics (for gate oxides, gate spacers or device isolation). The description also contains information about the doping profile, such as the concentration of dopants, the species and the position of the pn junctions. Information about the placement of electrodes completes the input needed for the device simulation. In principle there are two ways to generate such an electronic plot of a semiconductor device:

1. Manual device set-up: For devices showing an electrical behaviour which is only of type 1D it is possible to generate the device by using the single doping profiles of the process technology obtained by SIMS experiments (Figure 4.1). Under the assumption that all the implanted specie is electrically active, the SIMS data are used directly or fitted then by analytical functions which are used in the mesh generator for device construction (Figure 4.2). This way of generating a device is very simple and provides fast results, and can be used in current CMOS technologies, e.g. for STI bounded diodes.

2. Process Simulation: a process simulator allows a device to be built up based on a process of record (POR). The single process steps of a process technology, such as diffusion or implantation, are modelled by the process simulator. By means of a so-called "mask file", which provides information about the lateral extension in the single process steps representing the function of a lithographic mask, a complete device containing information about geometry and doping profiles can be generated. Process simulators are available for 1D, 2D and 3D structure generation; among the most important simulators available are TSUPREM4™ (Synopsys Inc., Mountain View, CA), DIOS™ (ISE AG, Zurich, Switzerland) and ATHENA™ (Silvaco International, Santa

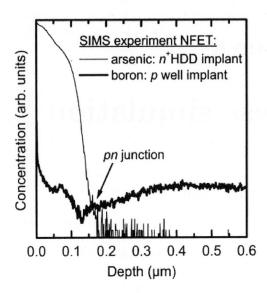

Figure 4.1: SIMS experiment resolving the vertical distributions of doping profiles of an
NFET device in a CMOS process technology. By superposing the single profiles, it is possible to set-up a device manually.

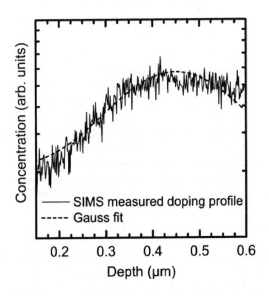

Figure 4.2: Reproduction of a measured doping profile using analytical functions. The
characteristic curve parameters like position of maximum and standard deviation are implant specific (specie/dose/energy).

Clara, CA). Compared to the manual set-up of devices, 2D or 3D process simulation provides a much more realistic description of the device, such as is often needed for example to properly predict breakdown voltages of *pn* junctions formed by lateral diffusions (Sze, 1985).

In the following, the most important process steps like implantation, oxidation and diffusion are discussed in more detail. In particular, modelling and implementation in a process simulator are highlighted. For practical use, some basic guidelines are given for how to calibrate the various model parameters. Process options which are relevant to ESD, but are not properly covered by process simulation are addressed in Section 4.4. The chapter ends with an example of how to build up an NFET by 2D process simulation.

For more details about processing and process simulation, the reader is referred to Sze (1985) and Carey (1996).

4.2 The basics of the important process steps

4.2.1 Implantation

Ion implantation plays a key role in introducing dopant impurities into semiconductor devices during the device processing. Dopant impurity ions are accelerated to high energies (of the order 10–100 keV) before they impact on the semiconductor target and come to rest below the surface of the substrate. Thanks to precise control of the process conditions, ion implantation has replaced pre-deposition and subsequent drive-in as the most reliable and repeatable method of doping semiconductors. From the process modelling point of view, ion implantation provides the initial conditions for subsequent diffusion modelling.

For the process modelling of ion implantation in a process simulator, two different approaches are in use: (1) analytic distribution function models (Selberherr, 1984), and (2) the particle or so-called Monte Carlo models (Biersack, 1980).

Analytic distribution function: On the basis of analytical functions like Gaussian distributions or Pearson distributions, the corresponding moments of the distribution function such as expectation value (range), standard deviation, skewness and kurtosis, are fitted to a measured variation of dopant concentration with depth for a number of different impurities, different dose and energy values. The recorded values are saved in look-up tables used for profile construction by simulation (Figure 4.2).

Particle model: In the particle model approach, the implantation process is reconstructed at the atomic level. The energetic ion enters a solid crystalline or amorphous target and comes to rest below the surface by losing energy by two basic mechanisms, nuclear scattering and electron energy loss. The first process is a scattering of the nucleus of the incoming ion with the nucleus of an atom in the target. In each collision event the ion changes direction and loses energy. In the second mechanism, the ion interacts with the electrons

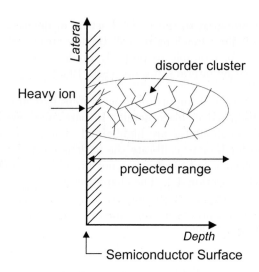

Figure 4.3: Example for an implantation damage by an impact of a heavy doping species
on a silicon target, after Sze (1985).

of the target atoms and slows down in "a manner of frictional drag" (Carey,
1996).

So long as experimental data are available to fit the moments of the distribution, the analytic approach gives very good results. Where such data are not
available for a given implant condition, it is necessary to use the more physically
based particle model to describe the ion implantation. On the one hand, particle
methods are very time consuming, since one has to track, at the atomic level,
each incoming ion plus the resultant damage tree as it travels through the bulk
(Figure 4.3). On the other hand the advantage of the particle method is that it
can model implantation into arbitrary geometries such as trench structures, and
can investigate the ion concentration profiles in lateral dimensions under a partly
covering mask layer. Additionally, channelling into crystalline materials can be
simulated very accurately by the particle model, which provides information on
the distribution of point defects due to crystal damage caused by ion impact.
These data are necessary for a subsequent high temperature diffusion simulation
(Hobler, 1988).

4.2.2   Diffusion

Diffusion of a substance from regions of high to low concentration is a basic physical
process. It is used in the processing of semiconductor wafers both to drive in
(pre-deposited) impurities, and to anneal damage caused during ion implantation.
The mechanisms behind the diffusion of dopants in semiconductors are highly
complex. Besides interchanging the positions of a dopant atom with an adjacent
silicon atom, it is thermodynamically highly favourable for the dopant atom to
bond with a vacancy or interstitial and then move through the lattice. Therefore

point defects or crystal imperfections strongly assist the diffusion process. The diffusion models in a process simulator account for the following effects (Carey, 1996):

Intrinsic diffusion: As long as the concentration of an impurity is below the intrinsic electron concentration of approximately 10^{18} cm^{-3} (at 1000 °C), the linear equation $\partial C/\partial t = D\nabla C$ is sufficient to model dopant diffusion. C is the concentration of the diffusing substance and the constant of proportionality D is called the "diffusivity". The diffusivity is temperature dependent and, as experimental studies show, follows an Arrhenius law.

Extrinsic diffusion: As the impurity concentration is increased so that the electron concentration n exceeds its intrinsic value n_i, the measured diffusivity shows a marked concentration dependence $D(C)$. A typical result is that the diffusion of boron is enhanced under the presence of a high boron background, while it is strongly retarded by a high background concentration of arsenic.

Field aided diffusion: Owing to the distribution of ionized impurities, the basic equations governing diffusion have to be extended by a drift term. This term can become especially important when two or more impurities are present.

Clustering: At concentrations near their solid solubility, impurities combine to form clusters consisting of several dopant atoms which are then effectively immobile.

Defects: In today's sub-micron technologies, the combination of high concentration diffusion steps and oxidation under low thermal budget processing leads to so-called anomalies in the diffusion profiles. These anomalies originate from the interaction of impurity atoms with host defects like vacancies, divacancies, self-interstitials and extended defects. For a reasonable representation of, say, the reverse short channel effect (RSCE) in MOSFETS (Rafferty, 1993), it is indispensable to consider such mechanisms.

Annealing process: Low thermal budget technology seeks to anneal damage while minimizing the diffusion redistribution of impurities. This is particularly important for the fabrication of very shallow junctions. Rapid thermal annealing (RTA) techniques are most successful in this regard.

4.2.3 Oxidation

The excellent dielectric properties of silicon dioxide, combined with its easy integration into silicon process technology, makes it *the* material of choice as an insulator (1) to achieve a separation of the single devices, and (2) to grow gate dielectrics. The two most important techniques for isolating neighbouring devices are trench isolation and the growth of a field oxide. Owing to the small trench width the first technique provides a high device density and has replaced field oxide isolation in today's VLSI technologies (Figure 4.4). Technically it is formed

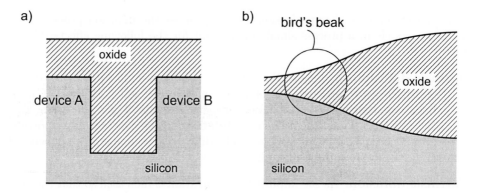

Figure 4.4: a) Shallow trench isolation (STI) and b) LOCOS isolation used to isolate
devices in integrated processes.

by chemically etching the substrate locally, and afterwards depositing an oxide
layer by a CVD process. The field oxide or LOCOS with its characteristic process
dependent bird's beak (Figure 4.5) is processed by partly covering silicon with
a hard mask like a structured silicon-nitride layer, and exposing the surface to
the oxidizing gas. With progressive oxidation, the silicon oxide layer grows and
deforms to the well known shape.

Alongside the isolating properties of the silicon dioxide, these layers have also
significant influence on device processing in total and on the electric characteristics
of the device. This makes it necessary to include a detailed representation of
oxidation modelling in the process simulator. This is especially valid in VLSI
technologies, where surface effects are becoming more and more important. As
mentioned in Section 4.2.2, impurities diffuse on a large scale via point defects.
Since oxide layers represent a large source or sink for point defects, they strongly
influence the redistribution of the impurities at the silicon oxide interface. This
requires accurate modelling of the oxidation process for a wide range of process
conditions, such as temperature, pressure and oxidation time.

The silicon oxidation problem is indeed very complex, since many different
parallel processes have to be considered (Carey, 1996): (1) diffusion of the oxidant through the silicon dioxide, (2) reaction at the silicon interface, (3) volume
expansion due to the reaction, (4) topology changes, (5) movement of the free and
reacting surfaces. The mechanical properties of the oxide layer depend strongly
on the processing temperatures, which are in the range 700–1200 °C. Oxide layers exposed to a steam atmosphere at temperatures above 1000 °C show viscous
properties leading to internal stress relief. For lower temperatures it is more appropriate to treat the oxide as an elastic material holding up strong internal forces
which influence the oxide layer growing process under an oxidizing atmosphere.
The stress modifies the diffusivity of the oxidant in the oxide, the reaction rate at
the silicon surface and the solubility of the oxidant in the oxide. The resulting oxide layer shape shows very process specific step coverage effects, such as a process
sensitive "birds beak" shape, or an oxide thinning at device corners (e.g. trench

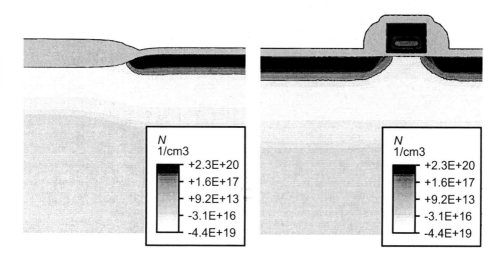

Figure 4.5: Left: LOCOS isolation showing the so called bird's beak. Right: Poly gate complex for an NFET device. At both positions, the resulting *pn* junction of drain to well shows comparable curvature (see colour section from page 269).

capacitors).

Especially the LOCOS formation process has a strong influence on the ESD behaviour of the device, because in many cases the LOCOS edge is the location of the lowest breakdown of the *pn* junction. As an example an NFET in a LOCOS technology is shown in Figure 4.5. The doping gradient and the curvature of the *pn* junction at the LOCOS and gate edge are comparable, indicating similar breakdown voltages.

4.3 Calibration of process technology parameters in a process simulator

The modelling parameters used for process simulation require careful adjustment in a physically reasonable range, in order to reproduce the structure correctly. Experimental data acquired chemically, electrically, and optically, are used as comparison and alignment with the "electronic copy".

The simulated doping profiles can be verified by SIMS experiments and electrical data. The SIMS measurements provide information about the vertical distribution of the doping profiles in the bulk. For a CMOS process, the doping profiles of interest for NFET or PFET devices are the well, channel, HDD (highly doped diffusion of drain/source) and LDD profiles. The simulated vertical profiles are compared to the measured doping profiles immediately after implantation or after a subsequent diffusion step. A recently developed two-dimensional dopant characterization method (Ukraintsev, 1999) now even allows the evaluation of the important lateral diffusion of dopants under the edge of an implant mask.

Since a SIMS measurement only resolves the chemical, not the electrically active concentration of impurities, it is also necessary to consider the electrical device characterization data for a proper calibration. For CMOS transistors the basic electrical parameters are leakage current, threshold voltage and saturation current. However, this calibration step requires the use of a device simulator. Extracting the different electrical parameters for a number of NFET and PFET devices with different gate lengths and at different bias conditions allows final calibration of the segregation coefficients for silicon/oxide and diffusion and damage parameters, separately for each implant type.

An inspection of the device cross-section is often performed to calibrate process parameters which determine the topology of a structure. As an example, to reproduce the birds beak shape of a field oxide, SEM cross-sectional pictures of manufactured devices are used to adjust those parameters, such as a stress-dependent growth rate, which govern the oxidizing process.

4.4   Limitations of process simulation of ESD elements

When investigating ESD protection elements on the basis of device simulation tools, one has to keep in mind that the process simulated devices incorporate some simplifications, which can have an effect on the electrical behaviour of the device under ESD stress conditions.

- Metal silicides: In most of today's applications, metal silicides like NiSi, CoSi or TiSi have become an integral element of the contact to lower the on-resistance of the device (Osburne, 1996). With the introduction of silicided diffusions, the ESD robustness of some standard snapback ESD protection devices has dropped significantly as a result of current filamentation (Amerasekera, 2002). To guarantee a homogenization of the current path along the entire device width, it is necessary to introduce a certain ballasting resistance, which is hardly possible with silicided diffusions because of the very low sheet resistance. Therefore, it is common practice in current CMOS technologies to partially block the silicidation. In general, the process simulators allow incorporation of the silicide formation process, including the partial consumption of the source/drain diffusion, during the growth of the silicide (Figure 4.13). However, it is not yet possible to reproduce the influence of stress induced to the silicon during the silicide formation process on the doping profile.

- 3D structures: The combination of lithography and lateral out-diffusion of implanted species leads to position-dependent junction shapes in real devices (see discussion in Section 5.2.1, Figure 5.5b)). The pn junctions of the drain and source diffusion along the gate edge have a cylindrical form, which turns spherical at the corners of the device. As the curvature of the spherical part leads to the highest electric field at the *pn* junction, the 3D nature strongly determines the behaviour of the device in the breakdown. However, because of the lack of experimental data it is very hard to calibrate and realistically model the 3D dopant distribution in the region of the device corners.

4.5 Example of a process simulation of NFETs

To illustrate the function of a process simulation, this section describes the generation of an NFET in a typical sub-quarter micron non-epi technology.

Definition of process steps

The set-up of the input file for the process simulation is structured like the process of record of the real process technology (as an example for the code of STI formation and the well/channel implant, see Figure 4.6). It includes the definition of

Deposition of layers: The deposition of material is described by a conformal coverage of a specified thickness. This describes chemical vapour deposition (CVD) very well, but is not applicable to an anisotropic deposition like sputtering. The process temperature is provided to account for the increase in the thermal budget.

Lithography: This process step is used to define a photo resist mask for the later etching, implantation or deposition step. The geometric boundaries are gained from mask files. These mask files are generated either manually or by an extraction from gds files. A negative or a positive photo resist can be chosen. The bias of the masks and the thickness of the resist have to be defined.

Etching: The various etch processes, such as wet etching, reactive ion etching, and plasma etching in semiconductor technology, are simplified to the removal of material, specified by isotropic or anisotropic behaviour and the etch depth.

Implantation of dopants: An implantation step is described by the species, dose, energy, tilt angle and rotation. As the implantation and later anneal leaves damages in the crystal like dislocations and point defects, which are very specific to the procedure and the implantation machine, the damage models have to be adjusted.

Oxidation: The oxidation is essentially influenced by the atmospheric parameters such as pressure, temperature, process gas (dry/wet oxidation) and the process time. At present, reliable oxidation models are available only above 600 °C. The mesh generation parameters are particularly critical for achieving a realistic result within a reasonable calculation time.

```
COMMENT #################### Example POR
########################

COMMENT #============ STARTING MATERIAL: P-TYP <100> 3.0 OHM-CM
==== initialize <100> boron=1e16 [...]

COMMENT #============ Shallow Trench Formation
==================== COMMENT #--------------- Oxide 10 nm
-----------------------------
    deposit    oxide                    thickness=1E-2    spaces=8
COMMENT #--------------- Nitride 200 nm
-----------------------------
    deposit    nitride                  thickness=200E-3    spaces=20
COMMENT #------------ RX-mask
---------------------------------------
    deposit    photoresist   positive   thickness=0.915    spaces=4
    expose     mask=RX
    develop
COMMENT #------------ IT etch
---------------------------------------
    etch       nitride       trapez     thickness=.300
    etch       oxide         trapez     thickness=.100
    etch       silicon       trapez     thickness=@STIDEPTH    angle=89
COMMENT #------------ STRIP RESIST
---------------------------------
    etch       photoresist all      [...]

COMMENT #============ P-Well Formation and NFET Vt Adjust
========== COMMENT #------------ BF-mask
---------------------------------------
    deposit    photoresist   positive   thickness=2.3    spaces=4
    expose     mask=BF
    develop
COMMENT #------------ P-WELL + N-CHANNEL implant
------------------
    implant    boron energy=600    dose=2.0E14    tilt=7    rota=0      +
    implant    boron energy=250    dose=1.0E13    tilt=7    rota=0      +
    implant    boron energy=65     dose=1.0E13    tilt=7    rota=0      +
COMMENT #------------ STRIP RESIST
---------------------------------
    etch       photoresist all      [...]
```

Figure 4.6: Example for a code of the first steps (STI and well/channel implant) of a
         typical POR of a CMOS technology. Typical process values are taken.

Diffusion and annealing: Annealing is a thermal treatment described by a temperature profile and an atmosphere. During the anneal the crystal structure is changed (e.g. recrystallization after an amorphization due to an implantation) and the dopants diffuse. In the latter case the segregation at surfaces and the solubility in the bulk has to be specified. Also processes like silicidation are described by an anneal after the deposition of a metal on the silicon surface. All other process steps including a thermal treatment, like oxidation and deposition, account for the diffusion by the definition of a thermal budget.

Construction of NFET

To build-up an NFET, there are typically seven major steps of the so-called front end of the process, which provide the most important parts of the device for the later ESD device simulation:

- Isolation of the devices by shallow trench isolation.

- Implantation of well and channel implants.

- Growth of the poly stack.

- Implantation of halo and/or pocket implants.

- Formation of the spacer.

- Implantation of the highly doped drain HDD.

- Silicidation.

The so-called backend process steps, including contact hole formation and the build up of the metal stack, may not be included into a detailed process simulation, but can be added in a geometric way, maybe deduced from a cross section.

Isolation

Starting material is a homogenously p^- doped wafer having a specific electrical conductivity. In modern CMOS processes the isolation of the different devices on the wafer is typically done by oxide filled into shallow trenches (Figure 4.7). The position and the area of the trenches is defined by a photo resist mask. At the position of an open resist mask, the trenches are etched into the substrate by dry etch processes. The depth of the trench varies with the technology, but is typically of the order of 0.4 μm. An oxide layer of about the same thickness is deposited and polished by CMP (chemical mechanical polishing). The oxide only remains in the trenches. Thus, a nicely controlled oxide stripe between neighbouring active regions can be generated by this so-called "damascene process". In practice there are deviations from the ideal process reproduced by the simulation, such as very narrow stripes arising from lithographical misalignment, and very wide STI stripes (such as under poly resistors) owing to a thinning of the oxide in the middle of a large STI area during the CMP process.

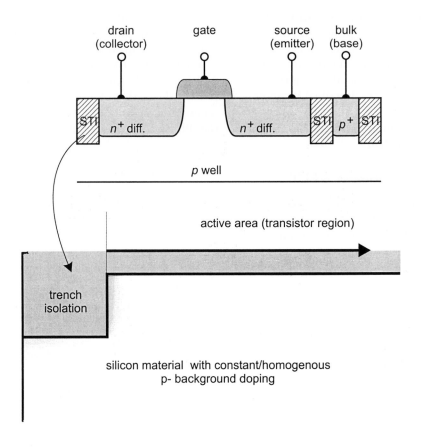

Figure 4.7: Lateral isolation of an active device by shallow trench isolation (see colour section from page 269).

Well

After the definition of the active area, the retrograde well profile is generated by high energy implantation of boron ions. Typically, this is a two or three step implant whose function is to optimize the drain–well capacitance, the breakdown voltage and the punch-through behaviour. In addition, a shallow p implant for the adjustment of the threshold voltage is also introduced using the mask of the well. A characteristic profile is shown in Figure 4.8. In more advanced technologies with retrograde wells, the well depth becomes lower and the sheet resistance increases. During the subsequent anneals, the boron close to oxide interfaces has a tendency to diffuse out of the silicon into the oxide. This leads to a typical boron depletion region, for example underneath the STI.

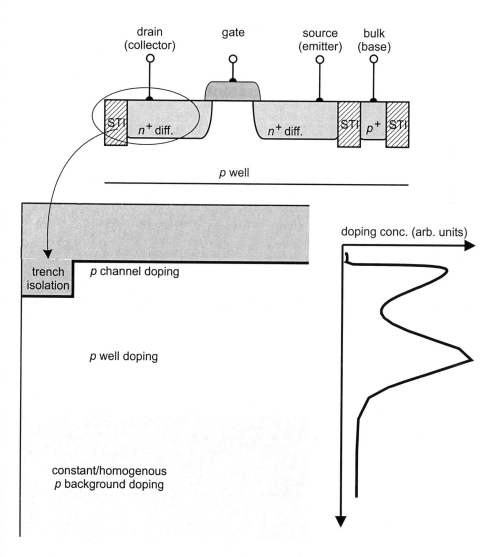

Figure 4.8: Implant of well and channel doping to control threshold and analogue behaviour (see colour section from page 269).

Poly stack

The formation of the poly stack of a standard logic process technology consists of the growth of the gate oxide, the deposition of the poly and the structuring of the poly by a plasma etch (Figure 4.9). The gate oxide is typically a dry oxidation resulting in an oxide layer of 1–5 nm depending on the technology and the device. 44 % of the layer has grown into the silicon eating up the substrate underneath. The poly layer is formed by a CVD process. The dopants of the poly can be inserted during the poly growth or by the subsequent doping together with the HDD formation. In the in-situ process the dopant is included into the gas providing the educt of the chemical vapour deposition. Technologically the major challenge is to minimize the depletion zone at the poly-to-oxide interface. This avoids an increase in electrical oxide thickness, which would degrade the performance of the transistor. This requires high doping in the poly close to the oxide interface which can be achieved best by in-situ doping. In future, however, other dielectrics and metal gate must be applied to meet the even higher performance requirements of CMOS technologies with gate lengths of a few ten nanometres.

LDD implant

The profile of the *pn* junctions at the gate regions of source and drain to the well have to be tailored properly to achieve the required breakdown and punch-through voltage and to minimize the negative influence of hot carrier stress for the oxide. *n* dopants like arsenic or phosphorous are used to form both thin doped layers close to the surface of 10^{17}–10^{18} cm^{-3} and deeper implants influencing the *pn* junction in the depth (Figure 4.10). To increase the punch-through voltage, an additional, shallow boron doping by large angle implantation has been introduced in quarter micron technologies and below. For devices connected to the pad and which are endangered by ESD, a so-called I/O LDD implantation is sometimes offered. For this implant, it is also important to get a properly calibrated implantation model.

Spacer

The spacer often consists of a thin silicon nitride layer of a few nanometres and a deposited oxide. Its function is the self-alignment of the subsequent HDD implantation to the gate. For this purpose the deposited oxide layer is etched back by a strongly anisotropic etch method like rapid ion etching (RIE). The nitride is used as the etch stop. Along the edges of the poly stack the oxide residue remains in a well defined thickness (Figure 4.11).

HDD implant

As one of the last process steps in the silicon, the highly doped implantation is performed. To minimize the on-resistance of the transistor a doping concentration as high as possible is needed. Usually the doping extends up to the solubility threshold of about 10^{21} cm^{-3} and the depth of the contact region is between

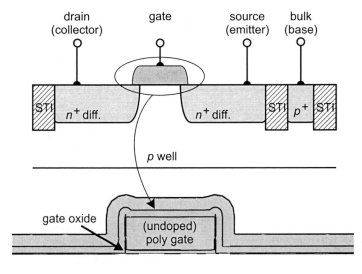

drain   gate   source   bulk
(collector)   (emitter) (base)

gate oxide

(undoped)
poly gate

p channel doping

p well doping

constant/homogenous p background doping

Figure 4.9: Deposition of poly silicon and structuring to form gate stack (see colour section from page 269).

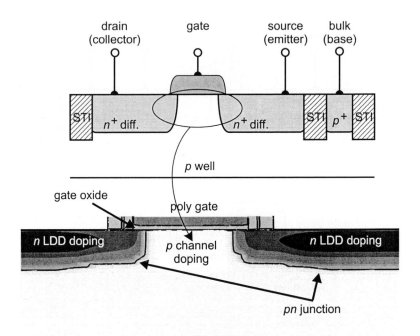

Figure 4.10: Implantation of source/drain extension (see colour section from page 269).

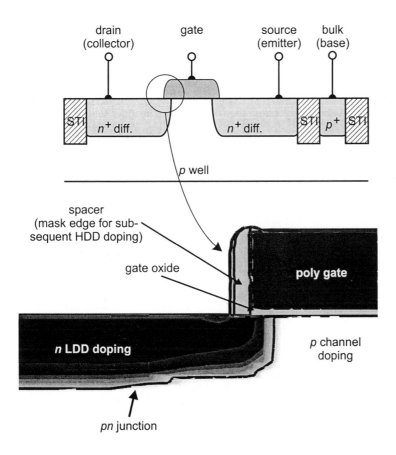

Figure 4.11: Deposition of poly silicon and structuring to form a gate stack (see colour section from page 269).

50 nm and 200 nm (Figure 4.12). Arsenic or antimony are used as dopants. The increase in the thermal budget arising from the annealing after the HDD implant also leads to a modification of the LDD and well profile.

Silicide

To further reduce the on-resistance very low ohmic silicide compounds are formed on the drain and source regions, mostly TiSi, CoSi or NiSi (see Figure 4.13). In this process step the metal is deposited on top of the silicon. Afterwards it is driven into the silicon by an annealing step where it forms the silicide. During the process the underlying Si is eaten up. Typically in the same process step the metal deposited on top of the gate reacts with the poly silicon and forms a layer of polycide. If the silicide formation should be blocked in certain areas where it is needed for ESD purposes, such as for driver transistors, the area without silicide growth must be covered by a hard mask (e.g. silicon nitride) before the metal deposition.

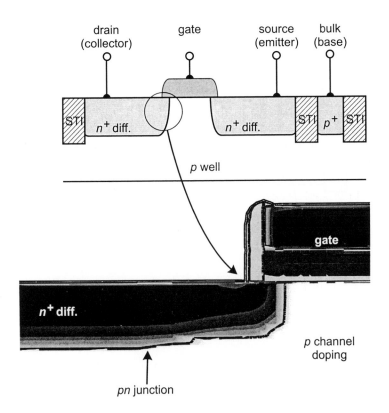

Figure 4.12: HDD implantation to reduce the parasitic diffusion resistance (see colour section from page 269).

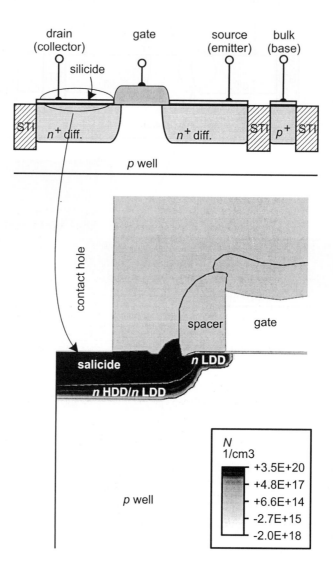

Figure 4.13: Silicide formation to reduce contact resistance (see colour section from page 269).

Summary

- Modern process simulators can reproduce most of the semiconductor technology process steps like oxidation, implantation and diffusion very well in 2D.

- Limitations affecting the later ESD device simulation are the reliability of 3D process modelling and the influence of silicide formation on the doping profile.

- For the later ESD device simulation it is important to have as exact as possible a doping profile, generated by the process simulator. The metallization and interconnects have less influence and can also be added in a geometric way without detailed simulation.

Bibliography

Biersack J.P.; Haggemark L.G., "A Monte Carlo Computer Program for the Transport of Energetic Ions in Amorphous Targets", Nuclear Instruments and Methods **174** (1980), 257.

Carey G.F., Richardson W.B., Reed C.S., Mulvaney B.J., *Circuit, Device and Process Simulation — Mathematical and Numerical Aspects,* John Wiley, New York, 1996.

Hobler G., Selberherr S., "Two-Dimensional Modeling of Ion Implantation Induced Point Defects", IEEE Trans. on CAD **7** (1988), 174.

Osburne C.M., Tsai J.Y., Sun J., "Active Elements of ULSI Contacts", J. Electronic Mat. **25** (1996), 1725.

Rafferty C.S., Vuong H.-H., Esraghi S.A., Giles M.D., Pinto M.R., Hillenius S.J., "Explanation of Reverse Short Channel Effect by Point Gradients"., Proc. IEDM (1993), 311.

Selberherr S., *Analysis and Simulation of Semiconductor Devices,* Springer, Wien, 1984.

Sze S.M., *Semiconductor Devices — Physics and Technology,* John Wiley, New York, 1985.

Ukraintsev V.A., Walsh S.T., Ashburn S.P., Machala C.F., Edwards H., Gray J.T., Joshi S., Woodall D., Chang M.-C., "Two Dimensional Dopant Characterization Using SIMS, SCS and TSUPREM4", Proc. IEDM (1999), 349.

Chapter 5

ESD device simulation

A device simulation is a virtual measurement which applies the same stimuli to a structure described by numerical methods as used in hardware measurements and observes the response of this "device under test". It has proven its value for the analysis of semiconductors for decades. Beside an early extraction of IV characteristics before hardware of the specific process technology is available and the fast assessment of possible process changes concerning their influence on the device behaviour, it also allows the insight into the physical processes happening in the semiconductor during the device operation. The inspection of the 2D or 3D current distribution and electric field often provides the development engineer the understanding how the device is working. While there is a long history of numerical calculations in the normal operational regime of CMOS devices for the investigation of the high current regime under ESD conditions one faces new challenges like the treatment of fast, transient lattice heating.

The first stumbling stone for a reliable device simulation meant to reproduce a real device behaviour, is the modelling of the critical device area as accurately as possible. For example, the junction doping profile of an NFET device underneath the gate essentially determines the low current characteristics, while usually the exact doping profile at the edge of the surrounding STI does not matter so much. Any error in the doping profile of the drain-to-gate edge influences the calculated IV characteristics far more than any physical transport parameter like mobility or lifetime which are commonly used to adjust simulation and measurement in the calibration procedure. That having been said, a complete and exact model of a real semiconductor device, even a simple transistor, is not possible (at least yet). As discussed in the previous chapter, the models, numerical methods and parameters used in 3D process modelling are not at all satisfactory, not to mention irregularities such as dislocations, segregation islands etc., which appear statistically in the device. Therefore in the set-up phase of the simulation the following issues have to be solved:

- Decide which part of the investigated structure will most likely contribute to the investigated effect. Particularly, does a possibly inhomogeneous behaviour along the width require 3D structures?

- Choose an appropriate mesh for the device simulation with a fine grid in the critical area, e.g. where high voltage differences or high density gradients might appear.

- Define appropriate electrical and thermal boundary conditions.

Reading this, it seems that one needs to know the result of the simulation even before the simulation has started. Unfortunately, this is correct – at least to a certain extent. To perform a simulation successfully, a picture or a model of the investigated effect has to exist in advance. The first task of the device simulation is to verify this principle picture. Only after passing this test the simulation can be used for the extraction of quantified statements. This "self-referential" feature of device simulation automatically leads to an iterative way of working. Principally, it is the foreknowledge of the engineer doing the simulation which determines how quickly the cycle of iterations converges to a satisfying status.

Approaching the problem of an ESD device simulation, a distinction has to be made between a low current and a high current regime. The borderline is often a matter of definition and depends on the type of device being investigated. Characteristically, in the low current regime lattice self-heating effects do not play a role for single, isolated devices. This is the typical operating regime of active devices like FETs. The cross-over into the high current regime of FETs is marked by the phenomenon of snapback, when the parasitic bipolar is triggered. Charge modulation, carrier recombination mechanisms and lattice self-heating are characteristic features which must be considered in the high current regime.

Before starting with the high current analysis and the incorporation of thermal effects, the parameters describing the transport at lower voltages and current densities like the FET mobility parameters have to be adjusted properly. This is done during the calibration phase of the simulation.

In the high current regime, the simulation should reproduce any snapback behaviour related to the trigger and sustaining point, the on-resistance in the high current regime, and the failure threshold. Clearly, the device simulator cannot simulate the melting of a silicon because of current filamentation, but appropriate thermal or electrical criteria can indicate a failure threshold. The correct treatment of these parameters requires the incorporation of electro-thermal coupling, i.e. lattice heating, and often 3D effects. These requirements complicate the numerical treatment and make convergence harder to achieve.

There are commercial general purpose device simulators on the market which have been applied for the purpose of an ESD device simulation, such as DESSIS™ (ISE AG, Zurich, Switzerland), see for example Esmark (2000); Stricker (2000); MEDICI™ (Synopsys Inc., Mountain View, CA), see for example Amerasekera (1994); and ATLAS™ (Silvaco International, Santa Clara, CA), see for example Boselli (2001). Any device simulator which is useful for ESD must at least contain electro-thermal coupling, flexibility concerning the topology of the investigated device, and the option to perform a 3D evaluation. The selection of the most appropriate tool should also consider the quality of the models, the flexibility of parameter adjustment during the calibration procedure, the numerical robustness (convergence behaviour, mesh generation) and the interface to the process simulator. The investigations described here are mainly done using DESSIS™ which provided sophisticated thermal models at that time.

In the following the basic equations for modelling of the electrical behaviour (Chapter 5.1) and their numerical implementation are discussed. The question, whether a 1D or 2D device simulation is sufficient to investigate reliably a certain ESD related problem or 3D device simulation is the tool of choice, is subject of the discussion in Chapter 5.2. There also the required boundary conditions are discussed. Before the numerical investigations can start, the physical models have to be calibrated to reflect the features of the considered technology (Chapter 5.3). Having done this, the derivation of a pre-silicon ESD protection concept for a ggNFET structure is performed (Chapter 5.4). Applying the device simulation on various other device types specific ESD relevant device features are highlighted (Chapter 5.5). Chapter 5.6 treats the influence of the ESD pulse shape for HBM and CDM pulses. The limitations of the device simulation for the analysis of the ESD behaviour is discussed in Chapter 5.7.

5.1 Basics of device physics and modelling

The main purpose of this chapter about the fundamental equations used in numerical simulation is to provide a qualitative understanding of the prerequisites and the constraints of the approaches. A derivation of the equations is not provided. Selberherr (1984) and Schenk (1998) describe in detail the models used for the device simulation. A more fundamental treatment of device physics is given in Seeger (1991). For an overview of the transport equations, Sze (1981) can be recommended. Thermal effects are also discussed in Ghandi (1978) and Baliga (1987).

Pure electrical transport is described by three systems of differential equations: (1) the Poisson equation, (2) the continuity equations, (3) the current equations. If lattice heating is included, in addition, (4) the heat flow equation must be considered.

Poisson equation: The Poisson equation, well known from the basic text books,

$$\vec{\nabla}(\epsilon \vec{E}) = -\rho \tag{5.1}$$

shows a relation between the charge ρ and the resulting electric field $\vec{E}$ adapted to typical semiconductor conditions. For this simple formulation a time-independent, homogenous permittivity ϵ has to be assumed. Also polarization effects arising from mechanical stress have to be neglected. For the standard materials presently used in the semiconductor technology, like silicon and silicon dioxide, these constraints are properly fulfilled.

Continuity equation: Also here an almost exact relation between the divergence of the current density $\vec{\nabla}\vec{J}$, the generation rate G, and the recombination rate

R of electron-hole pairs is given:

$$\vec{\nabla}(\vec{J_n}) - q\frac{\mathrm{d}n}{\mathrm{d}t} \;\; = \;\; q(G - R) \qquad (5.2)$$

$$\vec{\nabla}(\vec{J_p}) + q\frac{\mathrm{d}p}{\mathrm{d}t} \;\; = \;\; -q(G - R) \qquad (5.3)$$

with the electron charge $q = 1.6 \times 10^{-19}$ C and the electron (hole) density n (p). The only restriction concerns the influence of charged defects like deep recombination traps, dislocations etc., whose change in charge over time should be negligible. This relation is separately formulated for hole and electron currents. The equations must be solved for each point in space. In practice the challenge is to describe the recombination and generation rate correctly as a function of the local position and the current injection level.

Current equation: Here, by far, is the greatest simplification of an exact treatment. This equation describes the development of the investigated system over time. In principle, the starting relation for a system of classical transport, which already means neglecting quantum mechanical effects, would be the Boltzmann equation. It describes the motion of point-like particles, considering any interaction only during a short period of collision, treating each collision as Markoffian process. For the electrical transport it is extended to a semiclassical formulation by including quantum mechanical scattering rates and the band structure of the semiconductor. However, the Boltzmann equation is an integro-differential equation, which is not solvable by standard matrix diagonalization methods, but must treated by methods such as Monte Carlo (Jacoboni, 1989). This numerical approach is highly computationally intensive and can usually be applied only to rather simple device structures with little spatial variation of doping, topology, etc. Under certain circumstances, the Boltzmann equation can be simplified to a drift-diffusion (DD) equation by resolving the interaction integral analytically (Selberherr, 1984; Hänsch, 1991). The major constraints are

- only elastic scattering,

- weak spatial variation of band, structure, electric field and collision time compared to the mean free path,

- no degeneracy of the charge carriers and parabolic energy bands,

- lattice and carrier temperature in equilibrium and constant in the device (uniform, time independent),

- no velocity overshoot of the carriers,

- far from boundaries like contacts (several mean free paths),

- no Lorentz force.

Under these assumptions, the transport equations of the electrons and holes can be written as:

$$\overrightarrow{J_n^{DD}} = qn\mu_n\vec{E} + qD_n\vec{\nabla}n - q\mu_n n\frac{k_B T}{q}\vec{\nabla}(\ln(n_i)) \qquad (5.4)$$

$$\overrightarrow{J_p^{DD}} = qp\mu_p\vec{E} - qD_p\vec{\nabla}p + q\mu_p p\frac{k_B T}{q}\vec{\nabla}(\ln(n_i)) \qquad (5.5)$$

The equations relate the current density of electrons $(\overrightarrow{J_n^{DD}})$ and of holes $(\overrightarrow{J_p^{DD}})$ with the drift due to the electric field $\vec{E}$, via the mobility parameter μ_n (μ_p) and the diffusion by the diffusion constant D_n or D_p respectively. The last term including $\vec{\nabla}(\ln(n_i))$ accounts for any spatial variation of the intrinsic carrier density n_i. Often this term is included into the drift part by defining an effective electric field. k_B is Boltzmann's constant, $k_B = 1.38 \times 10^{-23}$ J/K.

However, this simplified version of the transport equation describes – by definition – neither the effect of significant lattice temperature gradients nor the acceleration of the charge carriers in strong electric fields, leading to a breakdown of the equilibrium between charge carriers and the lattice. There are several competing approaches to including large temperature gradients and high carrier energies consistently (Wachutka, 1991; Chen, 1993; Benvenuti, 1992; Pierantoni, 1993). Fortunately the situation is simplified for CMOS technologies with a gate length larger than 100 nm, where the heating of the carriers can largely be neglected and a reduced thermodynamic model is applicable (Wachutka, 1990):

$$\vec{J_n} = \overrightarrow{J_n^{DD}} - n\mu_n P_n \vec{\nabla}(T) \qquad (5.6)$$

$$\vec{J_p} = \overrightarrow{J_p^{DD}} - p\mu_p P_p \vec{\nabla}(T) \qquad (5.7)$$

where P_n and P_p are the thermoelectric power of electrons and holes respectively. This model provides a description of the heat generation and the influence of the gradient on the current distribution, which is essential for fast transients with high amplitudes like the ESD pulses.

Heat flow equation: In cases where the self-heating of the lattice is significant and the transients are neither in the regime of quasi-stationary conditions nor in the adiabatic regime, the heat flow must also be considered. Typically, for silicon devices, this situation is encountered during an HBM pulse with time constant of 150 ns. This requires the simultaneous solution of the heat flow equation:

$$C_p\frac{\partial T}{\partial t} - \nabla \cdot (\kappa \nabla T) = H \qquad (5.8)$$

where C_p is the specific heat of the material, H represents the dissipated heat and $\kappa(T)$ describes the thermal conductivity as function of the temperature. Heat flow and electric transport equations are coupled via the temperature T and the dissipated energy H as functions of the current density (Wachutka, 1990).

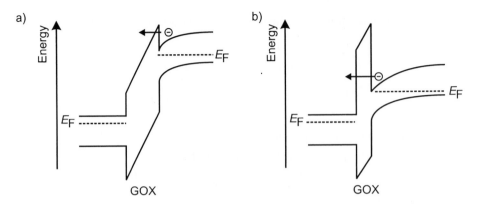

Figure 5.1: Schematic of a) the Fowler–Nordheim tunnelling mechanism and b) the direct
tunnelling. Band structure and Fermi energy are shown.

In many devices a further transport effect plays a role: the tunnelling through
dielectrics like the gate oxide. In modern CMOS technologies, it is a notable
feature even under normal operating conditions. For gate oxides with a physical
thickness of 1–1.5 nm the leakage of a MOSFET at least at room temperature
is essentially determined by the gate tunnel current (Ning, 2000). Two prominent tunnelling mechanisms through dielectrics, which have to be considered in
this concern, are Fowler–Nordheim (FN) tunnelling (see Figure 5.1 a)) and direct
(quantum mechanical) tunnelling (see Figure 5.1 b)). For higher oxide thickness
and higher potential drop across the insulator (like in EEPROMs) the tunnelling
is typically described in terms of the Fowler–Nordheim relation, while for thin
oxides, typically below 2 nm and comparatively low electric field in the insulator (< 6 MV/cm for silicon dioxide), the direct (quantum mechanical) tunnelling
governs the current.

Even after the set-up of the essential equations, no result can be gained without
a proper set of parameters. These can be divided into two groups with different
degrees of complexity. Some of them can be treated as rather invariant material
constants, like the permittivity ϵ of the semiconductor. Others, like the mobility μ,
the generation rate G, the recombination rate R, and the thermal conductivity κ,
are strong functions of position, electric field, temperature or process technology.
Their analytical relations are in many cases only a rough approximation, valid
only under very specific conditions. To achieve good agreement with experiments
in spite of the initial uncertainty concerning the correct parameters, it is necessary
to perform a "calibration" procedure before the high current device simulation is
started. This calibration has to be based on measurements which are specifically
sensitive for the variation of the considered parameter.

One of the basic parameters in the above equations is the charge carrier density.
Here, the device specific band structure plays the dominant role. Therefore, band
gap narrowing, such as from heavily doped regions, must carefully be incorporated.

Another group of parameters comprises the generation and recombination effects of excess charge carriers in the semiconductor device. The predominant

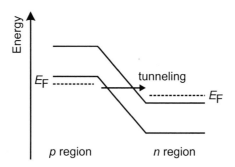

Figure 5.2: Schematic of the Zener tunnelling mechanism. Band structure and Fermi energy are shown.

effect at low electric fields and low current densities in silicon is the Shockley–Read–Hall (SRH) generation/recombination. The SRH mechanism describes the phonon assisted generation or recombination of free charge carriers via (deep) traps. The net generation/recombination rate is essentially described by the lifetime of the carriers τ, which is again related to the density of the traps. The device simulator distinguishes between the parameters for bulk and surface SRH generation/recombination.

In case of very high carrier densities, the so-called Auger recombination also plays a role. This is an interaction between three charged carriers where the excess energy of the recombining electron–hole pair is transferred to a third free charge carrier. A further physically distinct process of carrier generation is impact ionization. Here, high-energy charge carriers, which have been accelerated in high field regions, excite electrons from the valence band into the conduction band. This effect is typically the most important generation source during an ESD event, because of the extremely high fields appearing in the device. The impact ionization generation rate $G(\mathrm{II})$ is given by

$$G(\mathrm{II}) = \alpha J / q \qquad (5.9)$$

The ionization coefficient α is a strong function of the electric field. No strict relation is available, but several models are proposed which are usually offered in the device simulation (Overstraten, 1970; Okuto, 1975; Lackner, 1991). Another generation mechanism, which contributes to the reverse bias current of heavily doped pn junctions, is the band-to-band or Zener tunnelling which was already mentioned above. A transfer of charge carriers from the valence band to the conduction band by tunnelling at narrow pn junctions (Figure 5.2) leads to a typical shoulder in the IV characteristic below the onset of impact ionization at higher fields. Because of the more abrupt junctions and higher doping levels in modern technologies, this is a feature which has to be accounted for (see Figure 5.13).

To complete the discussion about the generation/recombination models, mention should be made of optical transitions, where the charge carriers are excited from the valence band by optical phonons. These only play a role for semiconductors with a direct band gap, like GaAs, and are negligible for silicon.

For the use of the transport model, the carrier mobility must be described. The various scattering mechanisms like acoustical phonon scattering, optical phonon scattering, impurity scattering, carrier–carrier scattering and interface scattering must be treated. The added effect of the various scattering phenomena can be described by a simple Matthiessen rule, just summing up all the different scattering rates.

To describe correctly the thermal influence of the transport, it is important to extract the thermal parameters thermal conductivity and heat capacity. For thermal conductivity, for example, the available relations provide good agreement only for a very limited temperature regime, for silicon, this is < 600 K for the default model in DESSIS (DESSIS, 1998). This is a problem considering the typical request for an ESD simulation with a maximum temperature close to the melting temperature of silicon at 1685 K. Therefore, the choice of the coefficients must be addressed in the calibration procedure of the simulation. Qualitatively, the thermal conductivity (e.g. of silicon) decreases with increasing temperature, while the heat capacity shows a weak increase.

The heat dissipation H causes another problem in the formulation of the thermal effects. A particular difficulty arises since lattice heating is not the only way of energy loss in the semiconductor. A consistent derivation of H is given by Wachutka (1990). This approach includes the influence of transient carrier density fluctuations, generation and recombination processes and the interaction with photons. In DESSIS™ (ISE AG, Zurich, Switzerland), the heat generation is assumed to consist of contributions from three sources: recombination, Joule, and Thomson/Peltier heating. Higher terms, including transient changes in the plasma density, are neglected.

After having determined the guiding differential equations and the required parameters, the system resp. device under investigation must be equipped with the appropriate electrical and thermal boundary conditions. In the device simulation, physical boundaries are partially given by the electrical contacts like source, drain, gate and bulk of a MOSFET. However, apart from these there are boundaries of the simulation domain which arise artificially from the limited extension of the simulated device. A wrong placement or conditioning of these boundaries might have a strong influence on the simulation results, and care has therefore to be taken to avoid this. The chosen boundary conditions must obey the physical laws like charge conservation. For example, for a current-controlled ohmic contact there are commonly accepted criteria, like the equivalence of the total current through the contact with the integral of the hole and electron current density across the boundary, thermal equilibrium and charge neutrality.

Finally the numerical treatment of the partial differential equations requires their reformulation in a way that is applicable for computer-based calculation (Selberherr, 1984; Bank, 1983). Currently there are mainly two approaches to deal with this problem. Neither provides exact results, but gives approximations to the original problem. In the first method of finite differences, the device is divided into sections along each spatial direction. For 2D simulation, this leads to a rectangular shaped mesh which is spread across the whole structure (Figure 5.3). The system of partial differential equations is then transferred into a system of difference equa-

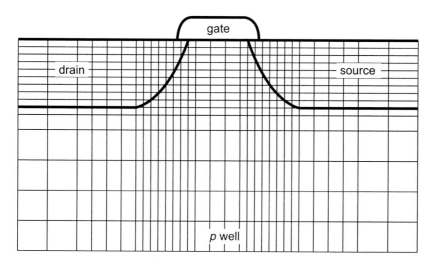

Figure 5.3: Draft of a rectangular mesh of a FET as used for the finite difference method.

tions. The second approach is usual referred to as finite element method. Here, for a 2D problem a triangular mesh is generated and for each of these triangular cells an approximation of the local solution of the partial differential equations is calculated (Figure 5.4). Of course, because of the inhomogeneous nature of semi-conductor devices, there are parts with little change in the parameters and others with extremely rapid change. The mesh building procedure has to account for this. The mesh must be adapted to find a compromise concerning the number of nodes between accuracy and calculation time.

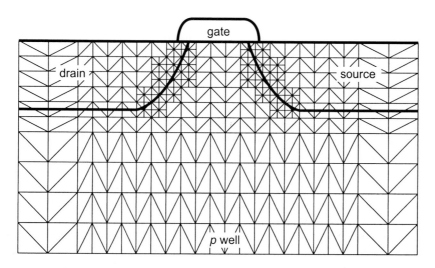

Figure 5.4: Draft of a triangular mesh of a FET as used for the finite element method.

5.2   Set-up procedure for ESD device simulation

Before starting now with the ESD device simulation a "practical" issue has to be solved. Even when a comprehensive methodology for 3D numerical solutions and elaborated models for electro-thermal coupling are available in the simulators, it might not always be advisable to use the full tool set to reduce the calculation time and to achieve better convergency. To support the choice of an optimum set-up of the simulation including the boundary conditions, the various options are discussed in the following.

5.2.1   The simulation approach: 1D, 2D or 3D simulation

The more dimensions that are taken into account in the simulation, the bigger are the numerical difficulties of the calculation. One problem is the computing time, which increases dramatically going from a 1D to a 2D or 3D calculation. Typical values of simulation runs for the different dimensions are given in Table 5.1.

Considering only the computing time for a 3D simulation run, it becomes clear that this is rather a tool for special case studies than to perform design studies with a large number of device variations – at least with the currently available hardware and software (solvers and algorithms).

Therefore it is often advantageous to use 1D or 2D simulations, which

1. qualitatively describe the specific device behaviour of interest,

2. can be used as a quantitative worst case/best case estimation for the real 3D device behaviour.

The 1D approach is useful for devices where the change of parameters occurs only along one direction, like for poly resistors at low current densities or forward biased junctions of vertical diodes. Also parts of more complex devices can be studied by 1D simulation, if one specific behaviour, like the breakdown voltage of a collector–base junction of a vertical *npn* transistor, is being considered.

Table 5.1: Computing times of some typical simulation problems for a workstation with a 360 MHz RISC processor.

Problem	Number of grid goints	Typical computing time
1D, electrical simulation of a breakdown voltage of a *pn* junction	10–50	0.5–3 minutes
2D, electro-thermal simulation of a TL pulse, transient	2.000–5.000	0.5–3 hours
3D, electro-thermal simulation of a TL pulse, transient	20.000–35.000	1–5 days

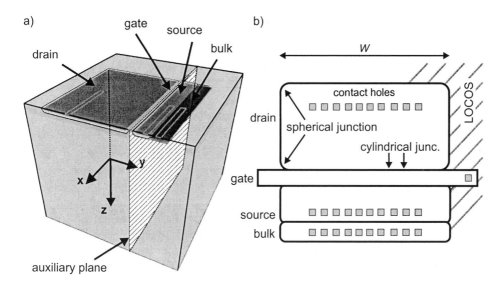

Figure 5.5: a) Generated NFET device for the 3D device simulation. The auxiliary plane is used to calculate the current flux under the gate of the transistor. b): Simplified layout of an NFET indicating the differently shaped junction regions (for process reasons).

The most common approach for devices like FETs and bipolar transistors is the 2D simulation. There, in general a 2D cut into the substrate in the yz-plane of Figure 5.5 is investigated. A prerequisite is that the parameter variations along the third space direction (x), e.g. along the width of the FET, do not govern the considered effect. This is usually fulfilled for the high current regime of wide FETs, when a certain current density is reached and before filamentation starts.

If this is not the case, a 3D simulation is inevitable. This typically occurs for devices driven into breakdown or into snapback. 3D behaviour also dominates when studying the mechanisms at the ESD failure threshold. As discussed in Section 4.4, reliable input from a 3D process simulation is often not available. In this case, the 3D device structure must be constructed from 2D process simulation. The approach is to translate the 2D doping profiles of the different parts of the structure, which can be determined very precisely, into a set of analytical functions. The 3D mesh generator uses these analytical functions to assemble the 3D structure (Figure 5.5a)), taking into account a suitable description of the lateral diffusion of the doping profiles for the third dimension (x axis). Devices defined in this way incorporate structural inhomogeneities along the perimeter of the structure (spherical/cylindrical junctions, (Figure 5.5b)) but neglect statistical inhomogeneities arising from non-ideal processing. These include roughness of the gate poly or other fluctuations of the gate length, which automatically vary the doping distribution in the silicon underneath.

To save computing time for 3D simulations, it is recommended that some simplifications are introduced.

1. For symmetry reasons, the simulation can be restricted to half of the device width.

2. For thermal worst case considerations, additional layers of the dielectric on top of the silicon can be neglected assuming adiabatic boundary conditions.

5.2.2   Boundary conditions for device simulation

Alongside the specification of the physical models, another focus for ESD device simulation is the set-up of electrical and thermal boundary conditions and the choice of a proper size for the finite simulation domain. The reason for this is that the device behaviour should not be influenced by artificial boundaries at any time during the simulation. Of course, this is in direct conflict with the attempt to keep the simulation area and grid as small and coarse as possible to save computing time. Therefore, one is forced to trade off between numerical effort and accuracy.

Electrical boundary conditions

The electrical boundaries are defined as equipotential lines or planes at the edges of the simulated structure. In the ideal case they match with the physical contacts, such as the contact holes. Care has to be taken with any artificial potential profiles introduced as a result of the placement of "simulation related" contacts which do not coincide with real contacts. For example, the depth of the simulated area should be large enough to avoid disruption of the device behaviour from the backside electrode (see Figure 5.6). The electrical conditions at these boundaries are given by the circuit connected to the contacts which models the high current discharge. The waveform and the equivalent lumped element representations for the different ESD discharge models have been discussed in Chapter 1. Taking the example of a ggNFET, source, gate, and the p well contacts are tied to ground and the drain is forced according to the considered discharge model. In the following, consideration is mainly given to HBM and TLP stimuli. Typically, the backside electrode is floating.

Another parameter that influences the choice of the size of the simulation domain is the charge carrier distribution, which is also determined by carrier diffusion. For all the investigated technologies described here, the ggNFET has been realized in a bulk p substrate. Because of this configuration, carriers that have been injected during an ESD event into the p substrate can deeply diffuse into the substrate before they recombine. There is no further barrier in the form of a pn junction or buried layer that can stop the penetration of these carriers.

For a TL pulse of 150 ns, the charge carrier density injected into the substrate (Figure 5.7) has been plotted along a vertical cutline starting at the silicon surface and crossing drain diffusion, p well and p substrate (see the cutline depicted in Figure 5.6). Charge carrier modulation results in regions where the density of the plasma strongly exceeds the background doping concentration of the bulk. Since the electric field in the substrate is negligible, the motion of the carriers is determined by diffusion. Over time, this causes an accumulation of carriers at the

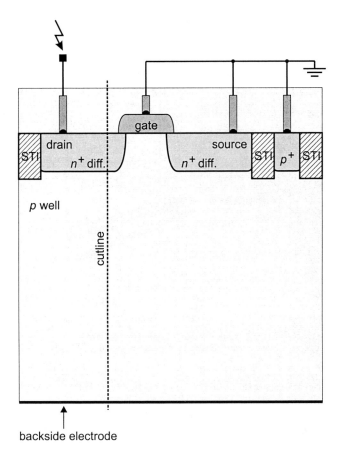

Figure 5.6: Definition of a ggNFET for device simulation. Gate, source, and bulk contact are shorted and grounded, while the ESD stimulus is forced into the drain contact. The cutline is used to estimated the necessary depth of the simulation domain.

bottom of the simulation structure. The characteristic depth of the penetration can be estimated by using the diffusion length $L_{n,p}$ (Sze, 1981) of the carriers

$$L_{n,p} = \sqrt{D_{n,p}\tau_{n,p}} \quad \text{with} \quad D_{n,p} = \frac{k_B T}{q}\mu_{n,p} \tag{5.10}$$

Values for carrier lifetimes $\tau_{n,p}$ and mobilities $\mu_{n,p}$ can be taken from a simulation test run. In this case the diffusion length is in the range of 15–20 µm (see Figure 5.7). To find the appropriate depth of the structure, a simulation pre-run is performed with a coarse grid and a depth of the structure of $\approx 3 \times L_{n,p}$. The profile of the diffused charge carriers (electrons *and* holes) in the substrate is compared to the profile resulting from a simulation with reduced depth (here: $1 \times L_{n,p}$). If both profiles are in reasonable agreement, the reduced value can be used for the final simulation using a refined grid. Especially an artificial crowding of carriers at the bottom of the simulation structure has to be avoided.

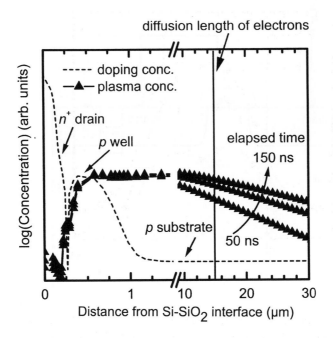

Figure 5.7: Doping concentration and plasma density for the ggNFET along the cutline defined in Figure 5.6 during a 150 ns rectangular current pulse. The p
substrate is completely charge carrier modulated.

Thermal boundary conditions

The power dissipated by the protection element during an ESD event leads to an
increase in the lattice temperature. As explained in Section 5.1, this demands a
self-consistent solution of the equations, including the electrical transport equation
and the heat transfer.

To guarantee a realistic heat propagation inside the structure, the thermal
boundaries are determined in a similar manner to the electrical boundaries. Obviously, the final dimensions of the device are given by the maximum of the extensions resulting from electrical and thermal boundary conditions. The counterpart
of the charge carrier diffusion length $L_{n,p}$ is the thermal diffusion length L_{th}. For
silicon it is about 3 µm for a time scale of 100 ns which is relevant for HBM events
(Esmark, 2000). For characteristic dimensions of ESD relevant devices, this leads
to the fundamental conclusion that during an ESD event no thermal equilibrium
is established inside the structure which is significantly larger than 3 µm. This
assumption can also be verified by BLI experiments. A typical result from BLI is
presented in Figure 5.8. Here it is demonstrated that the temperature continuously
increases during a 150 ns long rectangular pulse. This non-equilibrium condition
can only be described by a transient simulation; a quasi-static simulation is not

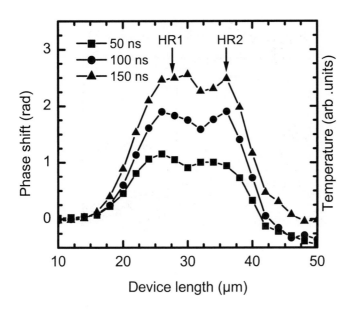

Figure 5.8: Measured phase shift of an ESD protection element at different times during a TL pulse along the length of the device (Esmark, 2000). Two distinct hot spots HR1 and HR2 are detected. With elapsed time, the temperature increases. The measurement was done by BLI.

able to cope with such a situation.

It is worthwhile to have a look at a cross-section of a real ggNFET for modelling the thermal boundaries (Figure 5.9). Besides the drain, source, and bulk contacts and the position of the gate, the complete system of isolation (inter-metal oxide, IMOX) and interconnecting layers (metal 1, 2) as well as contact holes and vias are shown. All these different layers must be considered for the definition of reasonable boundary conditions. This is either done by solving directly the heat equation in an extended volume including the dielectrics, the contacts and metals, or by a more abstract approach using a thermal network connected to isothermal planes of the device as shown in Figure 5.10 a). However, the usefulness of thermal networks is limited because of the missing implementation of thermal capacitances C_{th} into the existing simulators. This would cause a problem for the dielectrics. Therefore a compromise is favoured, whereby the contacts and their embedding dielectric layer are included in the simulation area. The thermal resistances are then placed on top of the contacts representing the connecting metal (Figure 5.10 b)). Considering the structure and geometry of the metal layers into which the heat flows, the thermal resistances R_{th} can be calculated on the basis of the results presented in Hirsch (1993).

An example of the temperature distribution in the ggNFET under TLP stress is shown in Figure 5.11. The hot spot, which is located at the gate edge of the drain diffusion, has a temperature of about 600 K after 150 ns and radiates heat

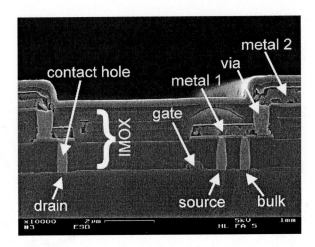

Figure 5.9: SEM picture of a ggNFET designed in a 0.35 µm technology.

in all directions. Owing to the reduced thermal conductivity of the IMOX, the isothermal lines are much closer compared to the underlying silicon. The excellent heat conduction across the contact hole becomes visible.

5.3   Calibration of device simulation

5.3.1   Basic considerations

Proceeding in the ESD device simulation flow, the selection of the simulation models must be performed and the set of parameters has to be assigned that describe the device behaviour under ESD stress.

For the CMOS process technologies discussed here with gate length larger than 0.1 µm, the modelling of the carrier transport can be based on the reduced thermodynamic model (see Equation 5.6), including the Poisson equation and the continuity equations for electrons and holes. Taking into account the kind of physical mechanisms important for a FET operating under high current injection conditions, one can specify the physical models in the device simulator according to the proposal of Slotboom (1993).

Mobility: Low doped regions will become carrier modulated under high current injection, leading to the need to incorporate carrier–carrier scattering in the mobility model. The forward bias voltage drop of bipolar power devices operating at high current densities is determined by this process (Baliga, 1987).

Generation/recombination: The current gain of a parasitic bipolar transistor operating inside an NFET device is a two carrier problem and the amount of carrier recombination finally determines the snapback of the device characteristic. Excess carrier recombination can happen, for example, via the

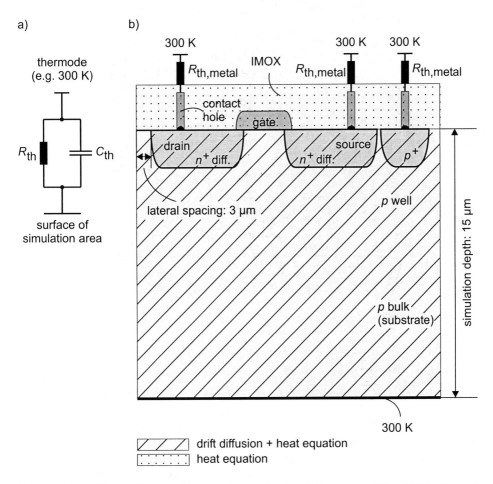

a)

thermode
(e.g. 300 K)

R_{th} C_{th}

surface of
simulation area

b)

300 K   IMOX   300 K   300 K

$R_{th,metal}$ $R_{th,metal}$ $R_{th,metal}$

contact
hole   gate

drain   source
n^+ diff. n^+ diff. p^+

p well

lateral spacing: 3 μm

p bulk
(substrate)

simulation depth: 15 μm

300 K

▨▨ drift diffusion + heat equation
⬚ heat equation

Figure 5.10: a) Thermal network to terminate the simulation area with suitable thermal boundary conditions. In principle, heat capacitance C_{th} and thermal resistance R_{th} can reflect the thermal influence of the surrounding material in an abstract way, but unfortunately a definition of a C_{th} is not supported by the software. b) Final definition of the thermal boundary conditions by means of thermodes (300 K) and thermal resistances R_{th} for the ggNFET. The depth of the simulation area of 15 μm is prescribed here by the electrical boundary conditions, while a lateral spacing of 3 μm is necessary because of thermal boundary conditions.

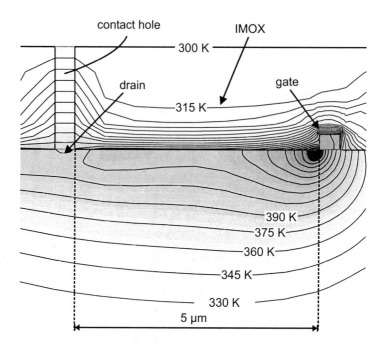

Figure 5.11: Temperature distribution in a ggNMOS 150 ns after the beginning of a TL pulse of 8 mA/μm. The isothermal lines are plotted with a spacing of 15 K (see colour section from page 269).

Auger and SRH recombination mechanisms. Generation mechanisms like impact ionization play an important part in the description of the electrical breakdown of *pn*-junction.

In addition, the heat flow equation has to be considered. As a result of high current injection, the devices under investigation are heated up during ESD and might be damaged due to thermal overload. Therefore it is necessary to introduce heat generation into the set of physical models and to couple the temperature to the other electrical quantities within the reduced thermodynamic approach.

Experience shows that the default values in the physical models are usually appropriate and need only slight adjustment. If large modifications are required to get agreement with the experimental data, this most likely indicates an inaccuracy of the doping profiles being used. An exception can be made for the carrier lifetime within the model for SRH recombination, which is a largely technology-specific parameter and needs to be calibrated individually for each process. The role of the ionization rate as main generation mechanism during ESD and the calibration of the recombination lifetime is discussed below.

5.3.2   Process technology specific calibration

Impact ionization

In most of the technologies, the breakdown voltage V_{bd} of pn junctions in devices is caused by impact ionization. Impact ionization or avalanche generation is a fundamental charge generation mechanism and has to be explicitly specified in the device simulator.

According to the so-called "Lucky-Electron-Model" (Shockley, 1961), the ionization is initiated by a free carrier that has to be accelerated over a mean free path l in an electric field E. If it gains sufficient energy exceeding the threshold energy E_i, it is able to generate an electron–hole pair. The ionization rate α is given by

$$\alpha \propto \exp\left(\frac{-E_i}{qlE}\right). \tag{5.11}$$

As long as the variation of the electric field E over the avalanche region is smooth enough, the history of the impacting particle is not important and the use of such a local avalanche generation model is justified. Classically, a minimum threshold energy of $1.5 \times E_g$ (with the bandgap energy E_g) can be derived for the threshold energy E_i (Sze, 1981). However, experimentally determined values are very different (Maes, 1990). The most common relation for modelling avalanche generation is the empirical relation of Chynoweth (1958), which is related to the "Lucky-Electron-Model" Equation 5.11.

$$\alpha_{n,p} = \alpha_{n,p}^{\infty} \exp\left(\frac{-b_{n,p}}{E}\right). \tag{5.12}$$

The parameters $\alpha_{n,p}^{\infty}$ and $b_{n,p}$ have been experimentally determined in Overstraten (1970). Figure 5.12 compares the measured characteristics of the reverse biased drain-to-substrate junction of a ggNFET in a 0.35 µm technology, with calculated ones using relation 5.12. At about 9 V, the characteristic shows a sudden increase in the current, indicating the electrical breakdown of the structure. The measured breakdown voltage $V_{bd} \approx 9$ V is read from the IV curve at the arbitrarily chosen current level of 1 nA/µm and is reproduced quite well by the device simulation. However, there is a transition regime between 6 V and 8.5 V that can only be modelled if tunnelling effects are also taken into account.

Since the doping concentrations, as well as the abruptness of the pn junctions, are steadily increasing in modern technologies, the effects of tunnelling mechanisms are becoming a critical issue for the simulation of the breakdown voltage (see Figure 5.13). However, it has been shown that tunnelling has no significant influence on the high current characteristic of a ggNFET beyond the snapback (PARASITICS, 2000).

By applying the correct description of the temperature dependent rate for avalanche generation (Valdinoci, 1999; Esmark, 2001), it is possible to reproduce the value for the temperature dependence of the breakdown voltage dV_{bd}/dT for any pn junction by one single model. This is shown by comparing experimental and simulated values of a smart power technology and a 0.35 µm CMOS technology calculated by the same model (Figure 5.14).

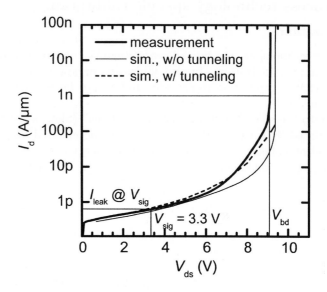

Figure 5.12: Measured and simulated reverse IV characteristics for the reverse biased
drain-to-substrate junction of a ggNFET device in a 0.35 μm technology
showing avalanche breakdown. V_{bd} is extracted at 1 nA/μm.

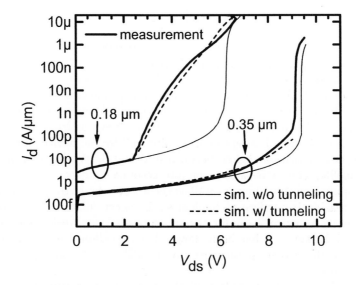

Figure 5.13: Comparison of the breakdown characteristics of a ggNFET in two different
CMOS technologies. Tunnelling plays an important role in the breakdown
regime of a 0.18 μm technology below 6V, while the IV characteristic is only
slightly affected by tunnelling processes in a 0.35 μm technology.

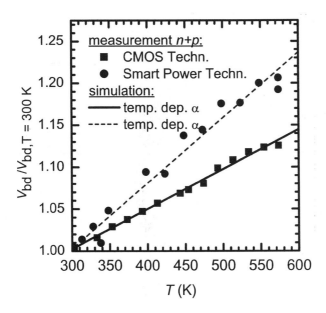

Figure 5.14: Modelling of the temperature dependent breakdown voltage in a smart power technology and in a CMOS technology. Owing to the different V_{bd} (approximately 10 V for CMOS and 40 V for smart power), the breakdown voltage has been normalized to its value at room temperature. For a reproduction of dV_{bd}/dT in both technologies, the same temperature dependent function α was used (Esmark, 2001).

Excess carrier recombination

There exist several mechanisms by which excess holes and electrons can recombine (Selberherr, 1984). They are classified into intrinsic (meaning an unavoidable material property) or extrinsic recombination processes. The class of intrinsic recombination processes covers effects like direct recombination (which can be neglected for indirect semiconductors like silicon) or Auger recombination (Häcker, 1994), while e.g. recombination via deep level traps (so-called SRH recombination) belongs to the set of extrinsic recombination process. Lifetime parameters characterize the single recombination processes, which are a measure of the probability of the particular recombination process to happen. For Auger recombination the coefficients that enter the description of the model are well accepted, so no further adjustment is necessary for this mechanism. Concerning SRH recombination, deep traps with an effective cross-section may originate by impurities or damage caused during ion implantation. As a result, the traps strongly depend on the type of process, and trap characterizing parameters have to be calibrated individually for the investigated technology. The empirical Scharfetter relation for SRH recombination links lifetime and the local background doping concentrations

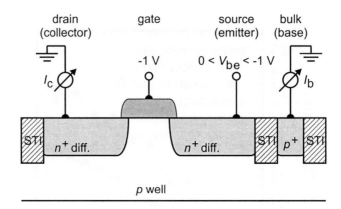

Figure 5.15: Cross-section and bias scheme of the device used to extract and calibrate process-specific lifetime parameters.

N (Selberherr, 1984)

$$\tau_{\text{SRH}} = \frac{\tau_{\max}}{1 + \left(\frac{N}{N_{\text{ref}}}\right)^{\gamma}} \qquad (5.13)$$

The quantities $\tau_{\max}$, N_{ref} and γ represent the independent fit parameters which are adjusted based upon the measured current gain (Figure 5.16).

One way to determine these parameters is to measure the current gain of a bipolar transistor. The current-dependent current gain of a bipolar transistor is sensitive to these recombination and lifetime parameters. Therefore, a simple four-terminal device (access to drain, gate, source, and bulk, see Figure 5.15) is suitable to perform this characterization. To suppress a MOSFET-current, the gate is held at a negative potential, in this case -1 V. The collector- and base-current are then measured as a function of the emitter-base voltage V_{be} and the current gain $\beta = I_c/I_b$ is calculated (Russ, 1999), see Figure 5.16. Finally, the measured bipolar current gain is used to adjust the excess carrier lifetime in the device simulator (Esmark, 2001).

In comparison to the default settings in the simulator, the value for $\tau_{\max}$ found in the example is relatively small ($\tau_{\max} = 6.5 \times 10^{-8}$ s), which is a consequence of the reduced thermal budget of modern CMOS-technologies. RTA technique, having a low thermal budget, allows the electrical activation of implanted species without large diffusion of the dopants. This behaviour is required for the construction of devices with shallow junctions and small dimensions. The reduced thermal budget does not allow a complete anneal of the implantation damage which leads to the reduced SRH lifetimes. Nevertheless, a good agreement between simulated and measured current gain can be achieved (Figure 5.16), especially in the range of high collector currents I_c which are important for ESD device simulation. However, since spatial information is missing in Equation 5.13, the calibration based upon this relation is rather an averaging over the entire device than a physically meaningful determination of a real lifetime, which is position dependent. This leads on the one hand to the observed deviation between simulated and measured

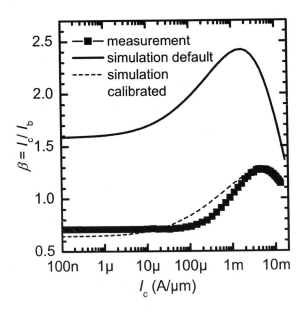

Figure 5.16: Measured and simulated current dependence of the bipolar gain of a bipolar transistor in a 0.35 µm process technology. The default parameter values in the simulation model for SRH recombination have to be adjusted to achieve good agreement with the experimental data.

current gain in Figure 5.16, and on the other hand to the consequence, that the calibration is device dependent. This averaging model is mostly justified for devices with one *pn* junction, where significant recombination processes occur. For SCR with at least two different junctions, this concept seems not to be applicable, which makes the approach of Equation 5.13 doubtful as an universal approach.

The input file of the device simulation must contain all these settings. An example for DESSIS™ (ISE AG, Zurich, Switzerland) input file for a TLP simulation of a ggNFET including the choice of models and device set-up is shown in Figure 5.17.

5.4 Derivation of a pre-Si ESD protection concept

As stated in Chapter 2, the strategy for a successful on-chip ESD protection concept is based on the idea of completely shielding the core region of an IC from an ESD event by providing a net of safe shunt paths in the I/O cell region. The I/O cell itself represents the interface from the external world to an IC and is a complex mixture of driver stages, input buffers and power bus lines. The target is to utilize this network to define a low ohmic electrical discharge path. Specific elements may have to be added to achieve this goal.

The concept of providing efficient ESD protection for an input or output cell

```
* Electrothermal Simulation of a ggNFET (0.13 um CMOS)
* under TL Pulse conditions: duration 150 ns, 5 mA/um

* specification of the electrodes, connect drain to a piecewise linear
* current source with 5 mA/um, 1 ns risetime and 150 ns duration
Electrode { { name="drain"  voltage=0
                             current= ((0,1.e-15)(1.e-9,5e-3)(150e-9,5e-3))}
            { name="source" voltage=0.0 }
            { name="bulk"   voltage=0.0 }
            { name="gate"   voltage=0   barrier=-0.55 } }

* specification of thermodes, pin metal connection to 300 K
Thermode { { name="thermtop"    temperature=300 }  }

* specify file names
File { grid   = "l_013_01_DCG1_mdr.grd"
       doping = "l_013_01_DCG1_mdr.dat"
       plot   = "id05_000_DCG1"
       output = "id05_000_DCG1"  }

* specify physical models
Physics { Recombination ( SRH(DopingDep) Auger
          Avalanche(vanOverstraeten))
          Mobility ( DopingDependence CarrierCarrierScattering
          HighFieldSaturation(GradQuasiFermi) )
          EffectiveIntrinsicDensity ( BenettWilson ) Thermodynamic }

* specify output values
plot { eDensity hDensity eCurrent hCurrent Totalcurrent
       ElectricField Potential Doping SpaceCharge
       SRH Auger Avalanche EffectiveIntrinsicDensity
       LatticeTemperature   }

* numerical settings
Math { Derivatives  Avalderivative  NoCheckTransientError
       NewDiscretization  RelErrControl  Digits=5 }

* specify equations to be solved
solve { Poisson
        Coupled { Poisson Electron Hole }
        Coupled { poisson electron hole  temperature }
        Transient ( InitialStep=5e-12  MaxStep=1e-8  MinStep=1e-16
                    Increment=1.2  InitialTime=0  FinalTime=150e-9
                    plot { range=(0, 150e-9) intervals=3}  )
                  { Coupled { poisson electron hole  temperature }}  }
```

Figure 5.17: Example of a DESSIS™ (ISE AG, Zurich, Switzerland) input file to simulate
a TL pulse of 150 ns duration and 5 mA/μm current density.

is shown in Figure 2.1. In the situation where the I/O cell is electrically stressed against one of the power supply pins VDD or VSS, the current which is forced into the I/O circuit must be shunted safely via a low ohmic path to the particular power rail. If this is not the case, then either the input buffer could be endangered by an electrical breakdown of the dielectric, or else the output driver could suffer a thermal breakdown.

One of the most straightforward approaches to an ESD protection concept is to equip the output driver itself with a sufficient ESD robustness. Such a self-protection concept avoids the adjustment to the *IV* characteristic of the output drivers caused by an additional ESD protection element. The conditions for a successful ESD protection concept can be reduced to a handful parameters (see Section 2.5). The derivation of these parameters on the basis of device simulation will be discussed in the following. Such an ESD protection concept extracted from the simulation will be called the "pre-silicon" (pre-Si) concept.

An important requirement for the realization of a self-protection concept is the scaling of the ESD robustness with the device width. In this case the required ESD robustness can be achieved by choosing an appropriate device width. Depending on the specific ESD requirements of the I/O cell as well as on the intrinsic ESD robustness of the FET, the devices can become very large. To optimize the area consumption, these devices are designed as multi-finger structures that might exhibit sequential turn-on of the single fingers under ESD stress (Amerasekera, 2002). Often one finger is triggered and initially conducts all the current until the current level I_{mf} is reached (see Figure 2.24). At this stage, the trigger voltage V_{t1} is exceeded again in the *IV* characteristics and a triggering of additional fingers is forced. If the width of the output driver is insufficient for a target ESD robustness, additional dummy fingers with their gates tied to ground can be inserted, which stay inactive during normal operation but become activated under ESD stress. To guarantee a homogeneous sharing of the ESD current between the active and dummy fingers, these dummy fingers should be designed according to the same layout rules as the fingers of the active device.

The concept of sequential triggering is violated if the failure current I_{t2} of the single finger is below multi-finger triggering current I_{mf} (Figure 2.24). Then the single triggered finger would be destroyed before the rest of the protection element has the chance to become activated. Neglecting any interaction between the single fingers the condition for a suitable multi-finger triggering can be rewritten in an idealized case as:

$$
\begin{aligned}
I_{mf} &< I_{t2} \\
\Rightarrow \frac{V_{t1} - V_h}{R_{diff}} &< I_{t2}
\end{aligned}
\tag{5.14}
$$

Two typical approaches to fulfill the requirement 5.14 are:

1. Provide a sufficient intrinsic ESD robustness I_{t2} of the device, by technology measures.

2. Reduce the current level for multi-finger triggering I_{mf} by layout measures.

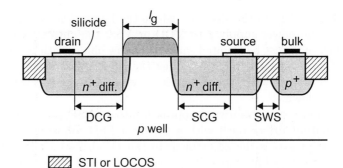

Figure 5.18: Cross-section of a standard NFET device with silicide blocked diffusions showing the ESD relevant layout parameters.

Table 5.2: Layout dimensions of an NFET in a 0.35 μm process used as the starting point for upcoming studies of design variations. The assignment of the different design parameters can be taken from Figure 5.18.

DCG	SCG	l_{g}	W	SWS
5.0 μm	0.7 μm	0.35 μm	100 μm	0 (ggNFET)

The holding voltage V_{h} the trigger voltage V_{t1} and the differential resistance R_{diff} can be modified by layout measures which are determined by the means of device simulation. However, it should not be forgotten, that the changes of the layout parameters also might also have an influence on I_{t2}.

In the following the deduction of the essential electrical parameters as function of the physical design parameters are shown following the IV characteristic of a ggNFET from the breakdown regime to the destruction level I_{t2} (see Figure 5.19). For the quantitative discussion, an example of a ggNFET processed in a 0.35 μm CMOS technology is used. The cross section of an NFET device including ESD design measures is depicted in Figure 5.18. Beside the influence of technological parameters on the electrical properties of the NFET, the high current IV characteristic of the device depends on its layout. The values for the drain contact-to-gate spacing (DCG), source contact-to-gate spacing (SCG), and source-to-well spacing (SWS) have to be chosen very carefully to optimize the ESD robustness and the clamping capabilities of the NFET output driver. The values of the most important design parameters of the "reference" device are summarized in Table 5.2. Essential features of the IV characteristic of this device can be gained by means of a 2D electrothermal simulation.

5.4.1 Breakdown regime

The breakdown regime of the ggNFET is governed by the avalanche multiplication at the drain–p well junction and typically exhibits in a very steep increase of the drain current by several orders of magnitude. This can be seen in the example

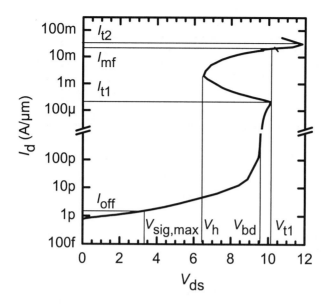

Figure 5.19: Simulated IV characteristics of the reference ggNFET in a 0.35 μm process technology. The parasitic bipolar transistor is triggered at a current I_{t1} indicated by the snapback in the characteristics to a sustaining point V_h. For even higher current values, the device shows an increase in voltage until the trigger voltage is V_{t1} is reached again at the current level I_{mf}. The device fails at I_{t2}.

at voltages $V_{ds} > 9.5$ V (see Figure 5.19). However, an increase of the leakage current can already be detected below the threshold because of Zener tunnelling (see Chapter 5.3.2). While for the ESD behaviour the breakdown voltage V_{bd} at low current densities is not relevant, it is important concerning the operating conditions of the protected circuit. Therefore it might be reasonable to define the breakdown voltage in terms of a maximum acceptable current density, also including lifetime considerations. Basically, the IV characteristics in the low-current breakdown regime of the ggNFET is determined by the doping profile at the gate edge and cannot be influenced by layout measures as long as punch-through between the drain and source region is avoided.

In the range above 100 μA the slope of the IV characteristic changes significantly. The ggNFET device is still operating in its reverse biased diode mode, but now the voltage drop across the substrate or well resistance is no longer negligible compared to the voltage drop across the reverse biased drain-to-p well junction.

5.4.2 Trigger point

The voltage drop across the well resistance leads to a forward bias of the source–well junction and finally activates the parasitic bipolar transistor. This condition

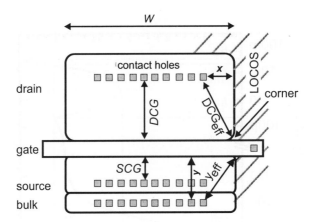

Figure 5.20: Top view of the ggNFET showing the arrangement of the different diffusions and the different effective design parameters.

is fulfilled when the avalanche generated current from the reverse biased drain–substrate junction reaches the threshold value I_{t1}. In the example, this occurs at a voltage $V_{t1} \approx 10.2$ V and a trigger current $I_{t1} \approx 0.2$ mA/µm. Subsequently, a region of negative differential resistance (called "snapback") appears in the *IV* characteristic. In general, the higher the effective p well resistance, the lower the required trigger current. The effective p well resistance of the protection element can be increased by process technology measures as well as by enhancing the distance y of the well contact to the gate. This has been studied by increasing SCG, which goes hand in hand with an increase of y in the chosen layout (Figure 5.20). The simulated trigger current drops with increasing SCG (Figure 5.21).

Using 2D simulation, the measured dependence of I_{t1} on SCG can be reproduced. However, especially for small values for SCG there is a higher deviation due to enhanced corner effects that can only be handled by 3D device simulation. The behaviour at the corners is mainly influenced by the parameters bulk resistance and avalanche generation.

1. Owing to the increased curvature of the spherical part of the *pn* junction in the corners of the drain diffusion, the breakdown voltage is reduced with respect to the cylindrical parts of the junction somewhere along the inner part of the transistor width. Thus, the most pronounced avalanche generation will appear there at low currents.

2. The effective p well resistance is increased for the corner regions of the device (Figure 5.20). To avoid destructive lateral breakdown at the field oxide edge, the contact holes are usually placed back a distance x from the sidewall of the junction. The trigger current I_{t1} is governed by the spacing y of the bulk contact to the substrate-to-source junction on the gate side. The value of the parameter is increased at the outer edges to y_{eff}. Owing to increased effective substrate resistance, the parasitic bipolar transistor is first activated in the corners of the device and is reduced in comparison to a 2D simulation run.

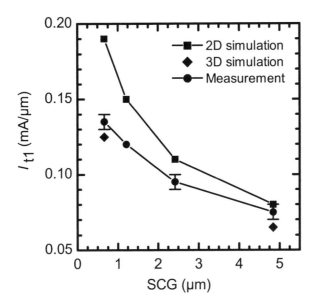

Figure 5.21: Trigger current I_{t1} as function of the source contact to gate spacing SCG. With increasing SCG the effective substrate resistance increases, leading to a reduction of the trigger current I_{t1}.

The ratio of y_{eff} to y, which is an indicator of the contribution of 3D effects to the overall device behaviour is highest for smallest SCG. This explains the growing discrepancy between 2D simulated and measured trigger currents for smaller values of SCG. The trigger currents calculated from 3D device simulation have been included in Figure 5.21 as well. There an almost perfect agreement to measured values is achieved.

Another result from the 2D device simulation is, for current levels above the triggering of the parasitic bipolar transistor, high current densities and high electric fields in the region of the drain-to-channel junction lead to remarkable power dissipation, causing a strong increase in local temperature. This necessitates the consideration of electro-thermal coupled transport. The evaluation in Figure 5.22 shows the simulated breakdown and snapback characteristic of Figure 5.19 again, now extended by the maximum temperature inside the device calculated for each point of the characteristic.

5.4.3 Holding point

The snapback of the IV characteristic after the triggering of the bipolar transistor leads to a reduced voltage at the holding (sustaining) point V_h of 6.5 V in the example (Figure 5.19). The magnitude of snapback to the holding voltage is ruled by the bipolar gain of the parasitic bipolar transistor. Under the assumption of

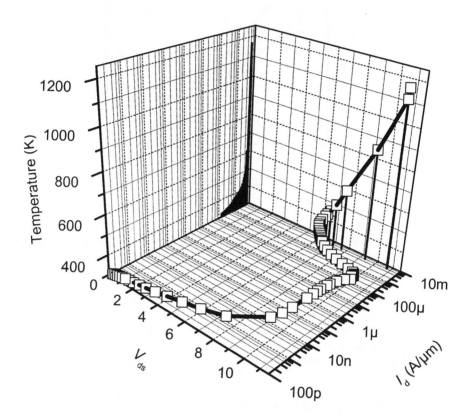

Figure 5.22: 3D representation of a device simulated IV characteristic for the ggNFET transistor. The projection of the curve to the I_d/V_{ds} plane shows the break-down and snapback characteristic already shown in Figure 5.19, while the projection to the I_d/T plane illustrates the maximum temperature inside the device for a given stress current value. Thermal effects play an important role beyond the trigger current of the parasitic bipolar transistor.

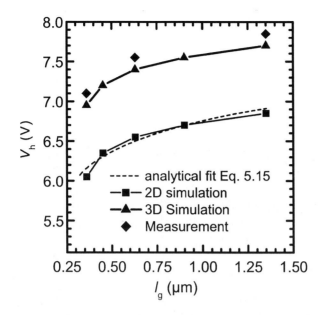

Figure 5.23: Holding voltage V_h for the ggNFET as a function of the gate length in a 0.35 µm technology. The sustaining point has been calculated additionally using the analytical expression 5.15 with fitted value for $n = 23$. The sustaining point V_h increases, as the gate length (equivalent to effective base width) increases.

a homogeneously doped base, the bipolar gain is inversely proportional to the Gummel number, which is the total number of impurities per unit area in the base region (Sze, 1981). Therefore, increasing the base width by extending the gate length l_g of the NFET device leads to a reduction of the snapback effect and an increased sustaining point V_h. This general behaviour is seen in the simulation result (Figure 5.23).

In the literature (e.g. Fong (1990)), there is an analytical model that correlates the holding voltage V_h to the breakdown voltage and the effective channel length $l_{g,eff}$

$$V_h = \frac{V_{bd}}{(\beta_{eff})^{\frac{1}{n}}} = \frac{V_{bd}}{k2^{\frac{1}{n}}} \left(\frac{l_{g,eff}}{L_n} \right)^{\frac{2}{n}} \tag{5.15}$$

where β_{eff} is the effective current gain of the lateral bipolar transistor, L_n is the electron diffusion length in the p substrate, and n is an empirical constant, $k = 1 \ldots 2$ is a geometry dependent factor. It is possible to fit Equation 5.15 to experimental or simulation data, but only with unreasonably high values for n ($n \approx 23$). According to the literature, reasonable values for n are in the range 2–6, depending on the shape of the junction. The large deviation between values for n extracted here and published in the literature is a clear indication that basic assumptions of the analytical models are no longer valid in the sub-µm regime.

For example, the prerequisite of a homogenously doped base is violated in modern sub-µm technologies.

In the preceding discussion it was implicitly assumed that the simulated effect (e.g. the current flow) is uniform along the width of the device. Only in this case 2D simulation is justified. However, in situations where, for example, the device is in an unstable state, like during snapback or second breakdown, there is a strong inhomogeneous component in the device behaviour, which requires a 3D treatment.

This is proven by, for example, a BLI analysis of a ggNFET in snapback (Figure 5.24 b)). For current densities below 5 mA/µm which affects the extraction of the holding point, the parasitic bipolar transistor is triggered only in certain parts of the total device width. The current level, at which a triggering of the full width of a (single finger) device, is achieved is defined as I_{fw}.

In this case 3D simulation is required. The first task of a 3D simulation is to model the triggering behaviour correctly. As with the previous example of a ggNFET in a 0.35 µm CMOS technology during triggering, there is a complicated current distribution along the width of the device depending on the current level. This can be reproduced very well by the 3D device simulation as shown in Figure 5.24. Depending on the waveform, only portions of the device either in the centre or in the corner are triggered below the full-width current I_{fw} (Esmark, 2001). One important consequence of this inhomogeneous triggering is an increase in the sustaining voltage and a pronounced kink in the IV characteristic when the current is increased further (Figure 5.25). This behaviour is explained by the triggering of only a limited width of the device (Figure 5.26), which causes a higher ballasting resistance and higher current density. The higher voltage drop along the diffusion region and a reduced current gain at the higher current density lead to the smaller snapback and a higher sustaining voltage. The current density does not change significantly while the triggered filament spreads along the width. This is reflected in the steep – almost vertical – IV characteristic until a kink occurs, when the device has triggered over the full width. The distinct shape of this IV characteristic can be reproduced by the 3D simulation, whereas the 2D simulation largely deviates from the measurements in this regime (Figure 5.25).

It must be stressed that the current inhomogeneities discussed in this section are not the consequence of a variation of the doping distribution along the gate width, or of the thickness of the gate oxide as a result of statistical process fluctuations. The profiles used here are smooth along the device width. Certainly, unintentional process fluctuations will further influence the current distribution, but these cannot be reproduced correctly in the doping profile.

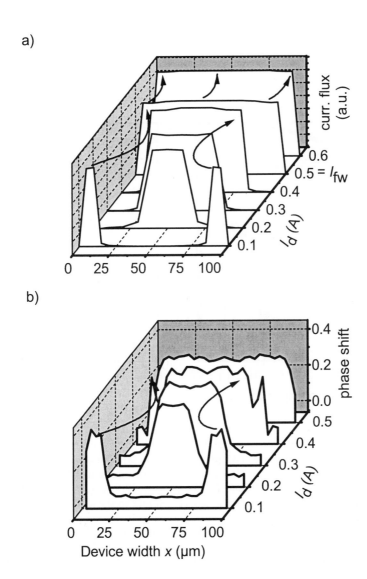

Figure 5.24: Current distribution of the reference ggNFET with $W = 100$ µm in a 0.35 µm technology along the width of the device for different TLP stress currents a) as predicted by 3D device simulation and b) measured by BLI. The value $x = 50$ µm corresponds to the middle of the device. For stress levels below 0.5 A, the parasitic bipolar transistor is turned on only at certain points along the width of the ggNFET.

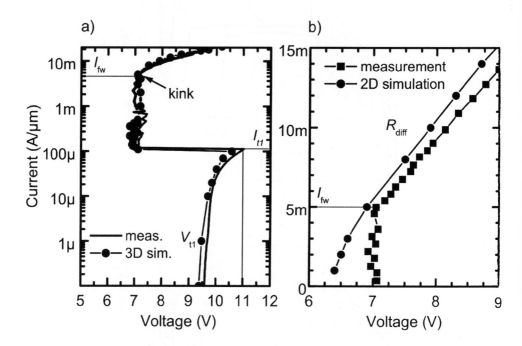

Figure 5.25: Measured and 2D/3D simulated high current IV characteristic for the reference ggNFET in a 0.35 µm technology. a) The 3D simulation is able to reproduce the kink in the characteristic at a current value I_{fw}. For the current levels between I_{t1} and I_{fw} the parasitic bipolar transistor is not triggered along the entire width of the device, leading to the characteristically vertical branch of the IV curve. b) The 2D device simulation predicts the parameter $V_h \approx 6$ V; however, the measured value is about 7 V. The trace of the IV curve in the region of the sustaining point is modified by some 3D effects (e.g. inhomogeneous triggering) which cannot be reproduced by 2D device simulation.

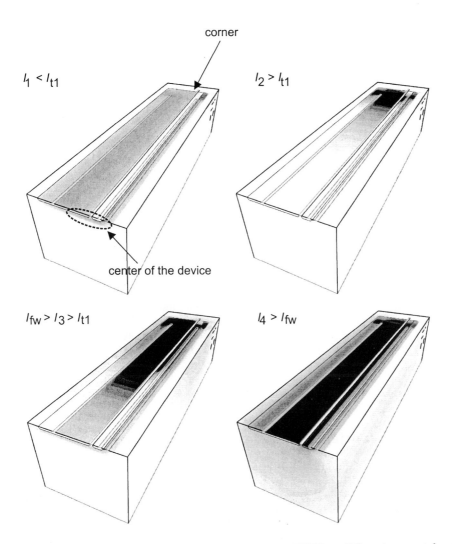

Figure 5.26: Current density distribution inside the ggNFET at different current levels. After snapback ($I_2 > I_{t1}$), the current is confined to a region at the device corners and spreads out for higher currents, until the entire device is triggered at I_{fw} (see colour section from page 269).

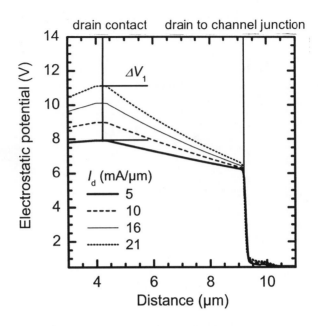

Figure 5.27: Analysis of the electrostatic potential distribution for the drain/channel region for different TLP stress levels. The main potential drop is across the
depletion region (at 9 μm) of the reverse biased drain–substrate (channel)
junction. A further potential drop builds up along the entire drain diffusion
(4 μm – 9 μm). With increasing I_d, the potential drop across the drain
resistance continuously increases, indicated by ΔV_1, giving rise to the differential resistance R_{diff} of the device while the potential drop across the
reverse biased drain to channel junction remains nearly unchanged.

5.4.4   Differential resistance in the high-current regime

Beyond the kink at the full-width triggering current I_{fw}, here at 5 mA/μm, a low-
ohmic branch is seen in the IV characteristic (see Figure 5.25), which is described
by a differential resistance R_{diff}. This differential resistance has a strong impact
on the ESD robustness of the device by acting as a ballasting resistance. It sup-
ports a more homogeneous distribution of the power dissipation along the device
width and influences the voltage clamping capability. At low current densities the
main voltage drop occurs at the reverse biased drain-to-substrate junction in the
gate region. However, with increasing current, the additional voltage drop along
the extended drain diffusion becomes comparable and determines the differential
resistance R_{diff} of the device (Figure 5.27).

It is common practice to modify the differential resistance in the high current
regime by an enhanced spacing for DCG or SCG, which leads to a higher voltage
drop in the diffusion region for the same current injection conditions (Figure 5.28).
Another positive effect of an increased contact-to-gate spacing is the reduction of

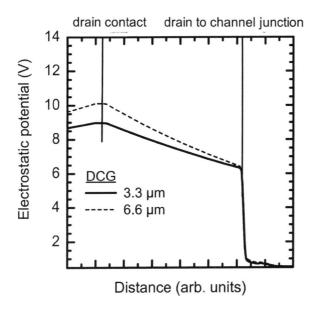

Figure 5.28: The potential drop across the depletion region of the reverse biased drain-to-substrate (channel) junction is independent of the design parameter DCGfor a specific TLP stress of 16 mA/µm.

the current level for multi-finger triggering I_{mf} that guarantees improved turn-on characteristics for all the fingers (Figure 5.29).

For investigating changes in the current level for multi-finger triggering we adopt the trigger voltage for the parasitic bipolar transistor from Figure 5.19, which is about 10.2 V. For the transistor with DCG = 6.6 µm, the current necessary to achieve the multi-finger triggering condition is about 16 mA/µm (Figure 5.29). This value increases for the transistor with DCG = 3.30 µm to 24 mA/µm. In other words, reducing DCG by 50 % requires an increase of the intrinsic robustness of the NFET by 50 % to achieve an equivalent multi-finger turn-on. If the value of I_{t2} in the NFET with DCG = 3.30 µm is lower than I_{mf}, the triggering condition for multi-finger structures would be violated and a reasonable and safe ESD concept could not be achieved. As shown here, for the device with smallest drain contact-to-gate spacing of 0.66 µm it is impossible to achieve the triggering condition again after the first snapback.

As both source and drain diffusion contribute to the differential resistance, evaluation of R_{diff} must take into account the sum of DCG and SCG. The rough analytic approximation for R_{diff} using the sheet resistance R_{sq} of the silicide-blocked source and drain diffusions commonly reads as:

$$R_{\mathrm{diff}} = \frac{\mathrm{SCG} + \mathrm{DCG}}{W} R_{\mathrm{sq}} \qquad (5.16)$$

The effect of an enhanced DCG and SCG on the differential resistance of the

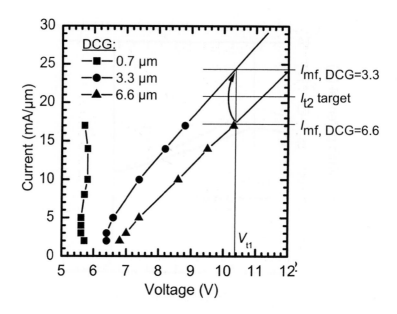

Figure 5.29: High current characteristic for the ggNFET device in a 0.35 μm CMOS
technology for different contact-to-drain spacing DCG. The current level for
multi-finger triggering is depicted for the different DCG.

silicide blocked ggNFET can be seen in Figure 5.30. In general, the differential
resistance continuously increases with larger contact spacing. In the case of the
0.35 μm CMOS technology, a slight saturation for large values of DCG may be
observed, as the result of an additional current path within the parasitic bipolar
transistor at large contact-to-gate spacing DCG (Esmark, 2001). However, as can
be seen in Figure 5.30, analytically calculated values deviate from measured values
for larger DCG, resulting in more optimistic (smaller) values if a certain R_{diff} has to
be achieved. In contrast, numerical simulation can nicely reproduce the observed
non-linear behaviour (Esmark, 2001). As the choice of these parameters is critical
for a self-protection concept, it seems inevitable that the choice of SCG and DCG
will be based on numerical simulations, so long as no measured values are available.

Despite these advantages, one has also to consider the disadvantages of an
increased contact-to-gate spacing:

1. an increase of area consumption, off current and junction capacitance of the
   driver,

2. a decline in the overall voltage clamping capability of the NFET driver.
   This must eventually be compensated by an increased width to meet the
   requirements of the ESD design window. However, this again worsens the
   aforementioned parasitics and the area consumption.

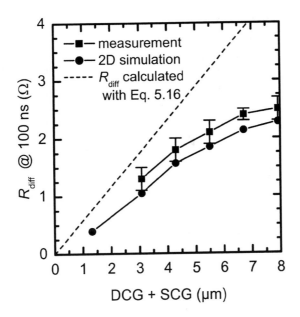

Figure 5.30: Differential resistance of the ggNFET extracted from measured and 2D simulated data as well as from approximation 5.16 for the 0.35 μm CMOS technology.

The total power dissipation of the device increases with increasing DCG, too. But since the voltage drop across the depletion region at the position of the hot spot, dominating the total power dissipation, is independent from the design parameter DCG, the ESD robustness per device width is also independent from DCG (Figure 5.28) as long as a certain minimum ballasting resistance is achieved. Finally, the increase of the resistance in the diffusion is limited by the vertical breakdown voltage of the drain diffusion–well junction. If the clamping voltage exceeds this value, the current along this path will destroy the device because of the missing ballasting resistance for this vertical breakdown. In choosing an optimum value for DCG, one has to find the best trade-off between the positive (ballasting resistance, multi-finger triggering) and negative (clamping capabilities, leakage, capacitance, area consumption) effects of a ballasting resistance.

5.4.5   ESD threshold simulation

The remaining parameter for deriving a pre-Si ESD protection concept is the current-to-failure I_{t2}. There are many reasons for the irreversible breakdown of semiconductor devices under ESD stress. They range from thermal overload of a specific region, to the shift of the threshold voltage of a transistor due to injection of hot carriers into the oxide, to the electrical overstress of isolation layers. With such a variety, it is clear that defining and continuously monitoring failure criteria during ESD simulation is a difficult task.

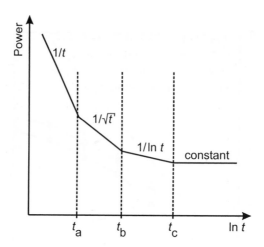

Figure 5.31: Power-to-failure distribution as function of the stress duration.

Hitherto, the analytical failure criteria to predict device breakdown discussed in
the literature (Wunsch and Bell, 1968; Abderhalden, 1990) cover only the thermal
overload of a specific volume inside a semiconductor device as a result of second
breakdown. The assumption is that a particular part of an ESD device heats up
under ESD stress as a result of large power dissipation. The particular part can be,
say, the reverse biased drain–channel junction of an NFET device. Because of the
local temperature increase, second breakdown will be triggered in the device after
a specific time. As shown by Wunsch and Bell (1968), the decisive parameters
that determine the so called power-to-failure distribution of a structure are the
volume in which the power is dissipated, and the environment of the hot spot, i.e.
the heat conductivity properties of the surrounding materials. Depending on these
parameters, several time intervals t_{rmx} characterizing the development and status
of temperature increase and heat propagation can be defined (Figure 5.31). These
are the adiabatic time domain, the steady state regime and an intermediate range
where the dependence of the power-to-failure on the time follows a power law. One
problem of the approach of Wunsch–Bell correlating power and time-to-failure is
that one has to know the internal power dissipation in the region of the critical
hot spot, which is of course not possible without information about the internal
parameters such as is obtained during a device simulation.

Based on the Wunsch–Bell model, Abderhalden (1990) has developed a thermodynamic model, which proposes an analytical relationship for the time-to-failure
dependence of a ggNFET under ESD-like conditions such as a pulse duration of
several 100 ns. For sufficient extension of the drain (DCG > 2 μm), the current-
to-failure dependence for different pulse lengths t_{pulse} is given as

$$I_{t2} \propto (t_{\mathrm{pulse}})^{-\frac{1}{4}} \tag{5.17}$$

where the factor of proportionality is a mean value over several process technology
parameters (e.g. sheet resistance of drain diffusion) and physical parameters (e.g.

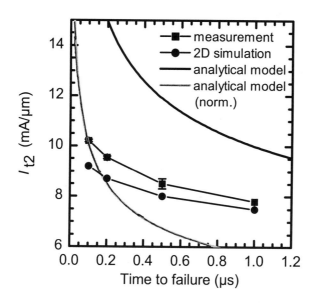

Figure 5.32: Measured current-to-failure for a NFET device as a function of pulse duration. The values predicted by device simulation and relation 5.17 are also
given.

heat conductivity, intrinsic temperature of local doping concentration).

To demonstrate the accuracy of such a model, we compare measured values
for I_{t2} as a function of pulse duration with device simulated values and values
predicted for I_{t2} using Equation 5.17. The comparison is made using a ggNFET
device in 0.13 μm CMOS technology, which fails due to second breakdown (and
therefore meets the fundamental requirements for the model of Wunsch and Bell).
Figure 5.32 shows the expected drop in the maximum acceptable current for the
NFET device for increasing pulse duration in the range of 100–1000 ns. In comparison, the values predicted by Abderhalden are higher than the measured values, in
other words they are too optimistic. For 100 ns pulses, the value is overestimated
by nearly 50 %. The prediction of the ESD device simulation is no more than
10 % away from measured values, and as shown in Esmark (2001), is always too
pessimistic. For development purposes this gives a much higher level of confidence,
as the device simulation results represent a kind of worst case estimation for I_{t2},
while the analytical model of Abderhalden gives, for an NFET, values that in one
process are too high and in another technology too low. Besides predicting absolute values for I_{t2} for a certain pulse duration, the Wunsch–Bell relation 5.17 can
be used to extrapolate from a short to a long pulse robustness. But even when
the Abderhalden distribution is normalized to the value measured for 100 ns, the
prediction for longer pulses is not as good as that using device simulation. For
example, the transition into the constant regime for I_{t2} at 1000 ns is not reflected
by the analytical model.

The reason why analytical models do not correctly predict second breakdown, is that a lot of parameters influence Equation 5.17, many of which are averaged values. This introduces a larger uncertainty, and without device simulation there is no approach to what is the relevant volume for estimating the trigger point for second breakdown.

In real world, a simple leakage criterion is often enough to detect a thermal breakdown or a drift in the devices performance as a result of an ESD stress. In contrast, in the simulation the detection of a fail is much more difficult. As no irreversible change of the structure is simulated, one has to use an indirect indicator like a temperature or electric field distribution. In addition, for the simulation failure criterion all of these different indicators have to be monitored and tracked at the same time. A proposal for a failure criterion, applicable for 2D ESD simulation is given in Esmark (2001):

$$I_{t2,sim}/W \equiv \quad \min\{ \quad J @ (T_{max} > T_{crit} \text{ locally}),$$
$$J @ (\text{occurrence of 2nd breakdown}),$$
$$J @ (E_{field,GOX} > E_{crit}) \ \} \tag{5.18}$$

It considered the minimum current density leading to one of the following effects:

1. Local excess of a critical temperature T_{crit} (like the melting temperature of silicon);

2. occurrence of an instability in the current, field and temperature distribution;

3. exceeding the critical electric field in the dielectric (like the gate oxide).

To conclude, the current-to-failure dependence obtained from 2D device simulation represents a powerful extension and improvement of the classical description using analytical expressions. The only analytical model published in the literature so far is linked to a certain class of device breakdown (second breakdown), but can not cover additional failure modes (for example, a critical temperature distribution of several hot spots in parallel). A fundamental advantage of the new methodology is that the criteria are independent of the device type and can be applied to any technology. In the following, there will be examples for each of the different contributions to the overall simulation failure criteria of Equation 5.18.

Critical temperature criterion, $T_{max} > T_{crit}$ locally

During an ESD event, the device internal power dissipation gives rise to a steadily increasing temperature distribution. Although short in time, the high current densities in combination with high electric fields can lead to temperature distributions and hot spots, which might cause changes in the material properties. Particularly when reaching the local melting point of a specific material, it is clear that irreversible changes in the device behaviour can occur, such as changes in the leakage behaviour. The simulator allows temperature distributions to be traced easily, and if a critical temperature T_{crit} is reached, the element can be assumed to have failed. The relevant temperature will depend on the position of the hot spot. A reasonable choice for the critical temperature T_{crit} is the melting point of silicon (1685 K), since the other materials used, such as Ti/W in the contact hole plugs, have comparable or even higher melting points (1998 K/3683 K). In older technologies, where the contact holes are AlSi plugs (melting point approximately 800 K), there has to be a more careful selection of T_{crit} for the different regions. Here, we discuss an example where the criterion of reaching a critical temperature distribution somewhere inside the device is triggered, thereby determining the ESD robustness of the element, for an NFET in 0.35 µm CMOS technology. Looking at the transient evolution of the maximum temperature inside the device when a stress current of $I_{TLP} = 26$ mA/µm is applied, one can see that the melting point of silicon is reached within 160 ns (Figure 5.33). The maximum temperature inside the structure is measured in a hot spot "HC" that is located within the drain diffusion directly beneath the contact hole. The heat source behind the hot spot radiates energy continuously into the surrounding material, but, in comparison with the heat generation, the heat removal is too slow. This non-equilibrium condition leads to a rapidly increasing temperature at the hot spot, until the melting point of silicon is reached. The hot spot below the contact holes dominates the maximum temperature in the structure throughout the entire pulse length and is both device and process technology specific. The higher the ESD current stress, the higher the voltage drop is along the drain diffusion. Depending on the process technology, it can happen that the breakdown voltage of the vertical pn junction below the contact holes is achieved, which leads to an opening of a current bypass into the substrate. In conjunction with the high electric fields in that region, a hot spot emerges. The result is that the parasitic bipolar is not only triggered at the drain–well junction at the gate (which leads to the hot spot "HG" indicating the activity of the lateral parasitic bipolar transistor of the NFET), but it is also triggered at the drain–well junction below the drain contact hole. This can cause difficulties, since this second parasitic bipolar transistor has no ballasting resistor, making the entire device unstable with respect to current filamentation.

The procedure for estimating ESD robustness on the base of a device simulation run is to search the device for a *single point* at which the critical temperature T_{crit}, e.g., the melting point of silicon, is exceeded. This criterion is surely too pessimistic, since it needs a finite volume, inside which the critical temperature of silicon is reached forcing a destructive phase transition from a solid to a fluid state. Nevertheless, in praxis it is the most useful approach, because the extraction of the maximum temperature in the device is much easier than to evaluate a complete thermal distribution over finite regions. For the evaluation of the ESD behaviour this delivers worst case results, which is preferable for the definition of a safe ESD protection concept.

Occurrence of second breakdown

The simulation based failure criterion for detecting second breakdown is also closely linked to temperature distributions, but in contrast to monitoring a material dependent critical temperature like the melting temperature, the intention here is to sense an (electrical) device instability leading to current filamentation.

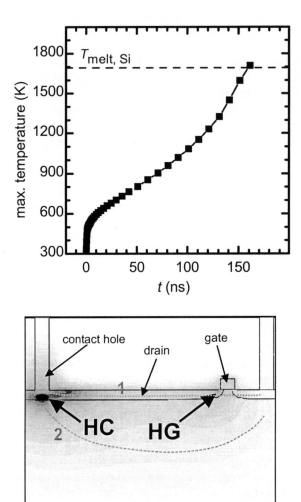

Figure 5.33: Transient evolution of the maximum temperature inside the NFET for a
current pulse of $I_{TLP} = 26$ mA/µm (top). The temperature exceeds the
melting point of silicon after 160 ns. The simulated temperature distribution for the NFET reveals a hot spot HC under the contact holes due to
a vertical electric breakdown of the drain-to-substrate pn junction at this
stress level (bottom). The hot spot HG is due to the laterally triggered parasitic bipolar transistor of the NFET and for the discussed stress condition
has a lower temperature during the whole pulse duration (see colour section
from page 269).

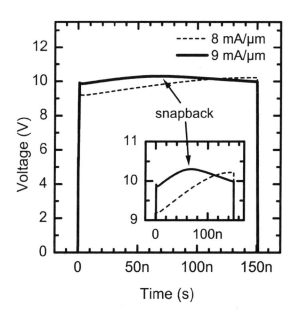

Figure 5.34: Simulated voltage waveform for a ggNFET in a 0.13 μm technology, with a current pulse of $I_{\mathrm{TLP}} = 8$ mA/μm and 9 mA/μm and 100 ns length. The second breakdown occurs for $I_{\mathrm{TLP}} = 9$ mA/μm after 70 ns.

Most devices showing second breakdown under ESD stress are irreversibly destroyed as a result of device instability. The term "second breakdown" is used to make a clear distinction from the first breakdown, which causes a reversible snapback of the device. As with snapback, second breakdown leads to a negative differential resistance mode of the device that is an indicator of instability in the device behaviour. Under second breakdown conditions, small perturbations lead to current filamentation, as shown earlier by the 3D device simulation for snapback. The situation is different for second breakdown, in that we are operating at much higher current levels, so the current density in the resulting filaments can cause hot spots with temperatures as high as the melting point of silicon, leading to damage as explained in the previous section. As a typical example the failure behaviour of a ggNFET in a 0.13 μm process technology caused by second breakdown is considered. The voltage waveform is shown for a square pulse causing a current density in the device of 8 and 9 mA/μm (Figure 5.34). In both cases, the voltage waveform exhibits first a transient increase, which is attributed to the decreasing mobility and rate of impact ionization as a consequence of the rising device temperature. While for the pulse with 8 mA/μm the waveform steadily increases till the end of the pulse, for 9 mA/μm after 70 ns a significant voltage drop of about 0.5 V is detected, which indicates second breakdown. This leads to a remarkable change in the slope of the resulting IV characteristic, which is extracted by averaging the current respectively voltage values in the time window

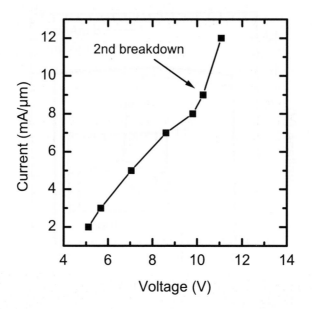

Figure 5.35: Measured and simulated *IV* characteristic obtained by TL pulsing (pulse
length: 100 ns). When the snapback arising from second breakdown is observed, the subsequent leakage characterization indicates damage to the device.

between 60 % and 95 % of the pulse length. This is shown in Figure 5.35.

The origins of the second breakdown are manifold. For ggNFET device it has
been shown (Amerasekera, 1994; Esmark, 2001) that the main reason for the triggering of second breakdown is the change in the composition of the base current
of the parasitic bipolar transistor from an avalanche dominated substrate current
to a thermally generated substrate current. For diodes or resistors, the reason for
second breakdown is the appearance of an intrinsic region in the device caused by
the local temperature increase (Amerasekera, 1994). The term "intrinsic" means
that the effective intrinsic carrier density exceeds the background doping concentration. Exceeding the intrinsic temperature should serve as a trigger for second
breakdown is that the device loses its original electrical properties, which are determined by the local doping profile. In any event, the second breakdown is triggered
by a temperature increase inside the device as a result of power dissipation.

GOX breakdown, $E_{\text{field}} > E_{\text{crit}}$

The clamping capabilities of ESD protection devices at relevant ESD stress currents must be kept low, to avoid destruction of the gate oxides and to ensure
protection of the internal circuit. The value for gate oxide breakdown depends
on the thickness of the dielectric as well as on the duration of the electrical stress
(Fong, 1987; Wu, 2000). Of course, the gate dielectrics of the protection elements

themselves are also subjected to an electrical overstress. This issue will become a severe problem in the near future, since the safe operating area, given both by the clamping characteristic and the voltage of GOX breakdown, is decreasing in the shrink path of future CMOS technologies. The resulting electric field strength depends on the layout parameters of the protection device, and it is therefore essential to make sure that the device design conforms with the requirements regarding GOX breakdown under ESD stress conditions.

The current distribution and the electric field in the gate oxide for an NFET device with minimum gate length in a 0.13 µm technology is shown in Figure 5.36

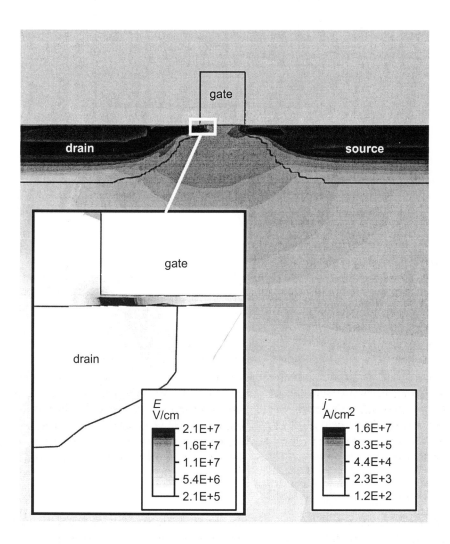

Figure 5.36: Current distribution and electric field in the gate oxide obtained by ESD device simulation for a 1.2 V NFET with $l_g = 0.13$ µm ($I_{TLP} = 8$ mA/µm).

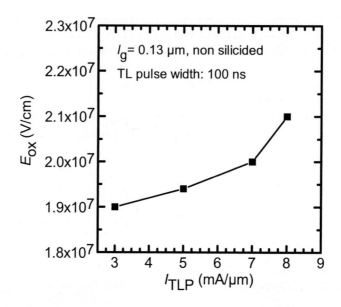

Figure 5.37: Maximum electric field in the gate oxide of a 1.2 V grounded-gate NFET as a function of the increasing stress current with 100 ns pulse duration.

(Salman, 2001). As demonstrated in the previous section, for this element the simulated I_{t2} at 100 ns is 8 mA/µm, since second breakdown occurs. The electric field is at its maximum in the region of the overlap of poly gate and drain diffusion (see inset in Figure 5.36). With increasing stress current I_{TLP}, the maximum electric field across the gate oxide rises steadily. In Figure 5.37, the dependence of the electric field is plotted as a function of the stress current. As expected, the risk of a breakdown of the gate oxide increases with higher stress currents.

In general, for a given stress current, devices with longer channel length show an enhanced electric field across the gate oxide, as shown in Figure 5.38. One reason for the increase in the electric field strength is the higher sustaining point. To detect a critical field strength in the gate oxide of an ESD protection element by means of simulation, one needs experimental data to serve as an upper limit. The measured time-to-breakdown of gate oxides as a function of the electric field is also plotted in Figure 5.38. This value was gained from experimental data in 0.13 µm and 0.18 µm process technologies, and is remarkably lower than the previously published values for critical electric fields in gate oxides by Wu (2000) for a PFET capacitor. Considering a transistor with 22 Å GOX thickness, the simulated electric field strength exceeds the experimental value for a critical oxide field strength if the gate length of the NFET is larger than 0.3 µm. As demonstrated in Salman (2001), devices with a larger gate length show a smaller ESD robustness than devices with minimum gate length, since the failure is no longer limited by

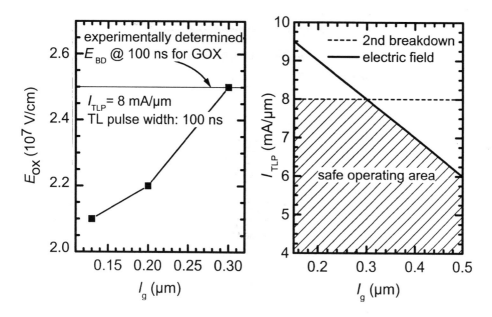

Figure 5.38: Electric field strength observed for 100 ns current pulses in gate oxide of NFET transistors of a 0.13 μm CMOS technology (2.2 nm GOX thickness) having different gate length. Depending on the gate length either second breakdown or dielectric breakdown leads to the fail. The safe operation area taking into account both effects is shown.

thermal overload but is governed by electrical breakdown of the gate oxide.

Oxide degradation

Beside thermal overstress and GOX breakdown, additional mechanisms exist that can lead to device failure as a result of an electrostatic discharge. An important mechanism is the injection of hot carriers into the gate dielectric during an ESD event (e.g. Groeseneken, 2000), since this can lead to shifts in the threshold voltage or transconductance of the device and, consequently, to a degradation of the circuit performance. Hot carrier injection effects are rather difficult to handle in the simulation, since they require exact knowledge about the energy distribution of the carriers in the bulk during ESD. Based upon a hydrodynamic calculation, it is at least qualitatively possible to specify the position of the injected hot carriers in the oxide and to understand their influence on the leakage behaviour, as shown by Pogany (2000). To estimate the position of injected carriers under ESD stress, the product of carrier temperature $T_{e,h}$ and carrier density $J_{e,h}$, $T_{e,h} \times j_{e,h}$, is considered as a measure of the carrier trapping probability into the oxide (Figure 5.39).

The simulation shows that $T_e \times j_e$ is bigger than $T_h \times j_h$, indicating that a higher injection of electrons can be expected. With the oxide locally negatively charged and a sufficiently high drain voltage V_d, the electric field at the Si/SiO$_2$ interface

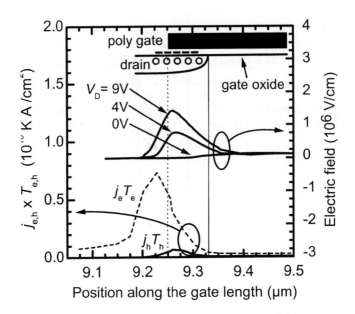

Figure 5.39: Simulated surface electric field and products of electron (e) and hole (h)
temperatures $T_{e,h}$ with current densities $J_{e,h}$ as a function of position at the
drain end of the channel for three different drain voltages of the reference
ggNFET in 0.35 µm technology. Negative injected charge (dashed line) in
the gate oxide is responsible for a very specific leakage behaviour (Pogany,
2000).

is high enough to cause defect-assisted tunnelling, giving rise to a very specific
kink behaviour in the leakage characteristic after ESD stress (Pogany, 2000).

3D behaviour causing device fails

In the discussion of the failure threshold it should not be forgotten that the physical failure is indeed a 3D effect and 2D device simulation gives only a rough
estimation. In the unstable state of snapback, the effective width of the triggered
parasitic bipolar transistor within an NFET device is the result of two competing
mechanisms: current spreading to minimize the voltage drop across the ballasting resistance, and current contraction to minimize the potential energy for the
avalanche source (Esmark, 2001). If the current contraction cannot be balanced
by the current spreading, a narrow filament is formed, which can easily destroy the
device through its very localized power dissipation. NFETs with an insufficient
ballasting resistance are candidates for such an early failure. This has also been
reproduced by 3D device simulation (Esmark, 2001); see Figure 5.40.

While for a NFET in a 0.35 µm process technology with DCG = 0.5 µm
the 2D device simulation predicts an I_{t2} of 10 mA/µm 3D device simulation and
measurement show an extremely low ESD robustness. As long as the ballasting

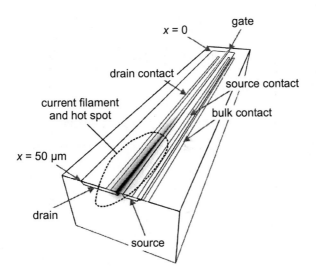

Figure 5.40: Temperature distribution for the ggNFET with minimum DCG showing a destructive hot spot in the middle of the device. The picture shows a I_{TLP} of 0.1 A. Because of insufficient ballasting resistance, the intrinsic current density in the filament for that stress level is too high, which causes high power dissipation in a confined volume, strong temperature increase and, finally, melting of the silicon (see colour section from page 269).

resistance is insufficient and current filamentation at low current densities occurs, 2D device simulation provides too optimistic results for I_{t2}. To be on the safe side, 3D simulation must be used to investigate the thermal breakdown of the filaments to the point where the critical temperature (or a critical field) is exceeded. A useful formulation of the failure criterion for 3D simulation is

$$I_{t2,3Dsim}/W \equiv \min\{ \quad J @ (T_{max} > T_{crit} \text{ locally}),$$
$$J @ (E_{field,GOX} > E_{crit}) \quad \} \qquad (5.19)$$

which reproduces the measured values more precisely. In comparison to the failure criteria for 2D device simulation, the condition of 2nd breakdown can be neglected for the 3D simulation since a current filament as a result of 2nd breakdown automatically fulfils the condition $T_{max} > T_{crit}$ (Esmark, 2001).

However, as the 3D simulation is numerically difficult, the detection of the second breakdown is seen as a reasonable approach to extract a failure threshold from a 2D simulation, if a uniform current distribution along the device width can be expected below the failure threshold.

5.4.6 Breakdown of parasitics

As already stated several times, the target of ESD development is to find ESD protection devices which meet the requirements of the ESD design window for

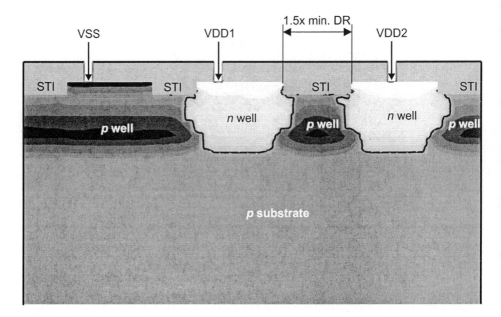

Figure 5.41: Cross section showing two n wells tied to different potentials VDD1 and
VDD2, representing a parasitic device with a specific breakdown voltage
between both power supply voltages (see colour section from page 269).

all possible stress combinations. A limitation of the ESD design window is the
maximum electric field strength in the GOX, which was introduced in the simulation failure criteria. However, there are also other voltage thresholds for critical
parasitic breakdowns which should not be exceeded during an ESD discharge.
The most important parasitic structure endangered by a destructive breakdown is
formed by two n wells, or n^+ diffusions, tied to different potentials (Figure 5.41).

In principle, such a critical parasitic can exist everywhere in the core of a chip,
which makes detection very difficult. The process technology only guarantees that
both n wells are isolated against each other under normal operating conditions, and
prescribes minimum design rules for this purpose. Under ESD stress, however, the
voltage difference between the two domains can be much higher. Because of this
high voltage difference, a punching between both n wells or even the triggering of
the parasitic bipolar can be observed as a result of insufficient lateral separation.
Many parasitic elements show a dependency in their breakdown voltage on the
distance, which allows the required breakdown conditions to be satisfied by an
appropriate choice of the design rules.

The simulation can now be used to determine:

- the upper limits of the breakdown voltages for such a parasitic;

- the minimum design rule for achieving the maximum breakdown voltage
  given by the junction breakdown to the well or to the substrate.

In Figure 5.42, the breakdown voltage of neighbouring n wells has been sim-

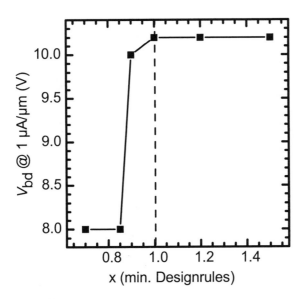

Figure 5.42: Device-simulated value for the breakdown voltage of the n well to n well parasitic as a function of the distance parameter. $x = 1$ denotes the minimum spacing according to the design rules of the corresponding technology. For $x < 1$ a significant drop of the breakdown voltage is observed which decreases the ESD design window.

ulated as a function of the lateral spacing. In this particular case, the optimum breakdown voltage was reached just by using the minimum design rule value for the spacing of n wells in the particular technology. However, it also shows, that e.g. a small misalignment of the masks or process fluctuations lead to a drop of the breakdown voltage. Therefore, an increased distance for ESD critical parasitics is recommended here. It can also be seen that the maximum achievable breakdown voltage is 10.2 V due to the breakdown to the substrate, which represents a layout-independent limit for the ESD design window.

5.4.7 Compilation of the information to a pre-Si ESD concept

Now with the gained information about the ESD parameter, an ESD pre-Si protection concept can be compiled, which meets the ESD design window. This can be fulfilled for a self-protection concept by an adjustment of the design parameters of the transistor. The dependence on the parameters has been discussed in the previous sections. Table 5.3 summarizes the observed tendencies.

The procedure of a successful derivation of a well operating ESD concept is demonstrated here for a 0.13 μm CMOS process technology.

First, a compromise between the ESD robustness and the clamping capability

Table 5.3: Effects of variations in the design parameters of an NFET device on the electrical parameters of the high current characteristic.

Parameter	tendency	
	positive	negative
DCG ↑	I_{t2} ↑, I_{mf} ↓	V_{clamp} ↑, I_{sat} ↓
SCG ↑	I_{t1}↓, I_{t2} ↑, I_{mf}↓	V_{clamp} ↑, I_{sat} ↓
l_g ↑	I_{leak} ↓, (V_h↑)	I_{t2} ↓, V_{clamp} ↑, I_{sat} ↓

on the one hand, and the leakage current on the other has to be found by adjusting the gate length l_g. From simulation runs the minimum gate length has been chosen as 0.15 µm to cope with the leakage requirements of the I/O cells at temperatures up to 125 °C without causing a significant drop in the ESD robustness (8 mA/µm, see Figure 5.43).

The trigger point can be influenced by the choice of the SWS distance. If a butted junction configuration is implemented for the driver, which is most area efficient, the important parameter is the distance SCG (Figure 5.21). The I_{t1} value decreases with increasing SCG. The exact level of I_{t1} is not of great importance for a self-protection concept. However, using, for example, a ggNFET as discrete

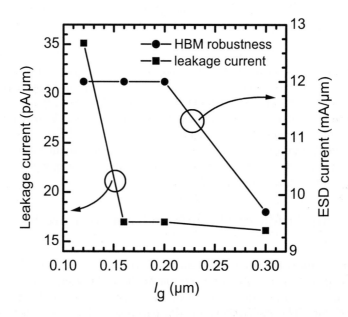

Figure 5.43: Leakage current of the NFET device of a 0.13 µm CMOS technology with different gate length. A slight increase in the minimum gate length causes the strongest reduction of the leakage current while the HBM robustness is nearly unaffected within the depicted gate length variation.

ESD protection element its trigger point must be below the triggering condition of the protected circuit, which can be achieved by appropriate placement of the well contacts of the driver and the protection element.

In the analysis it must be taken into account that measured (and 3D simulated) values of the trigger current I_{t1} are always below the current level that is predicted by 2D simulation, which usually represents a worst case condition for the ESD behaviour of the protection element.

With respect to an optimized ESD concept, the sustaining point of the snap-back device should be chosen as low as possible, assuming that it is sufficiently above the operational voltage. This measure guarantees the best clamping behaviour on the chip during an ESD event. This condition favours the shortest gate length possible in the given technology. But, as discussed above, the leakage/off current across the device increases with shorter gate length. Therefore, it can be more advantageous not to use minimum gate length, but instead adopt a somewhat larger gate length. In especially as the dependency of the holding voltage on the gate length is rather weak (see Figure 5.23). Another possible drawback of the lower holding voltage is that owing to the increased β of the underlying npn and, consequently, larger snapback, the device tends towards stronger current filamentation.

The 2D simulated sustaining point V_h is always equal (homogenous triggering) or below (inhomogeneous triggering) the measured values. This allows to reliably determine the danger of unintentional latch-up of the device under operating conditions. It also gives a worst case value for the choice of the ballasting resistor.

To ensure reliable multi-finger triggering after the snapback a sufficient ballasting resistance has to be added. This is commonly implemented by increasing the contact to gate spacing especially on the drain side DCG, which leads to a distributed resistor along the diffusion. In general, this region may not be silicided to achieve the necessary resistance value. On the other hand a too large spacing DCG could lead to a very high voltage drop in the diffusion region and finally to a break down of the diffusion to well junction just underneath the contact holes. This can also lead to an early fail. Beside this, the increase of DCG also increases the area consumption of the drivers and the protection elements significantly. The simulation allows to find the optimum for the drain-to-gate distance as shown in Figure 5.44. A value of 2 µm already ensures the multi-finger triggering of the ggNFET in the considered 0.13 µm CMOS technology. As after having reached the full width triggering current of a finger the 2D simulation reproduces the differential resistance very well, it can be used for this analysis with a good justification.

Up to the failure threshold 2D ESD device simulation results in a worst-case estimation and it can be used for the parameter extraction of a safe ESD protection concept. However, for the calculation of the failure threshold itself, there is the danger that the 2D simulation provides too optimistic values, because it neglects the current filamentation. Whether a 2D simulation can be used for the determination of the I_{t2} at all, strongly depends on the investigated device type and geometry. A ggNFET designed with a certain ballasting resistance or other measures to ensure uniform current distribution along the width of one finger be-

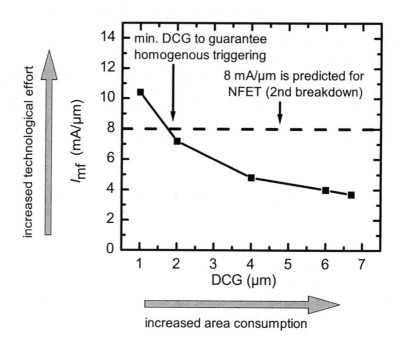

Figure 5.44: Calculated current level for multi-finger triggering as a function of the drain contact-to-gate spacing DCG for an NFET with gate length 0.15 µm. Based on this methodology, it is not only possible to design elements that fulfil the condition for a suitable triggering of multi-finger devices ($I_{t2} > I_{mf}$), but also to find protection elements with a minimum area consumption. In the particular case of a 0.13 µm technology, the 2D device simulation predicted ESD robustness was $I_{t2} = 8$ mA/µm and 12 V/µm under HBM stress. This was achieved by a DCG of 2 µm. If the desired ESD robustness has to be achieved with less area consumption, then further technological measures are required to increase the intrinsic ESD robustness of the process while still fulfilling the relation $I_{mf} < I_{t2}$ (since the current level for multi-finger triggering increases). To be on the safe side, the value for DCG chosen for the design of the first test chip is usually the optimum value proposed by the simulation, plus a safety margin.

low the failure threshold, can in general be treated by the 2D simulation (Esmark, 2000). Applying the criterion of second breakdown to the example of a ggNFET in the 0.13 µm CMOS technology an value of 8 mA/µmis gained which is in good agreement with the measured data (Figure 5.32).

Even if 3D simulation is applied for the extraction of the failure threshold there is some uncertainty because statistical fluctuation like gate length variation along the device width can not be properly included. Therefore an acceptable extra margin should always be taken into account for determination of the I_{t2} in a pre-silicon concept.

In conclusion this evaluation is very beneficial for the ESD engineer to optimize process technology and design at the same time. It provides reliable data for finding a trade-off between area consumption and technological effort. Increasing I_{t2} by means of process changes (e.g. junction depths) helps to reduce the minimum tolerable value for DCG, which leads directly to a reduction in area consumption for the protection element.

5.5   Device simulation of typical protection elements

Besides FETs, other ESD protection devices, such as diodes, resistors, vertical bipolar transistors and SCRs, are used for full-chip ESD protection concepts. This section presents some basic results for these devices in CMOS technologies or similar devices realized in other processes (e.g. BiCMOS, SOI) derived from ESD simulation, concerning in particular the influence of design and process parameters on the high current performance.

It is worth stressing again that the starting point and prerequisite for a successful investigation of devices under ESD conditions is a sufficiently accurate description of the devices structure, materials and doping profile. It also goes without saying that the device simulator has to be fully calibrated with respect to the process-dependent physical parameters.

5.5.1   Diodes

ESD protection concepts make wide use of diodes in forward bias as robust ESD protection elements. However, also the reverse bias modes of the diode is of interest for the ESD engineer. As the diode shows a very weak ESD performance under reverse bias stress, the focus here is on the breakdown voltage and maximum injection current. Under forward bias conditions, the parameters of interest are the clamping capabilities and the maximum injection current. Using the methodologies derived in Section 5.4.5, ESD device simulation can be applied to find the limits and to optimize the design. Another important parameter, accessible by TCAD tools, is the leakage current at elevated temperatures. For stacked diodes in particular, this is a critical parameter (see the discussion in Section 2.4.2).

If possible, the design and layout of a diode (Figure 5.45) should be optimized with respect to the electrical parameters mentioned above, but in some cases it is

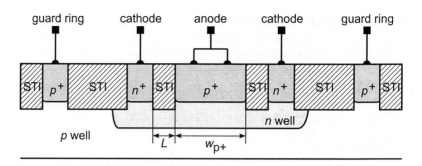

Figure 5.45: An STI-bounded p^+–n well diode in a CMOS process technology. The cross
section shows the diode designed in a double finger configuration, including
ESD-relevant design parameters. For latch-up purposes the entire diode is
surrounded by a p^+ guard ring. As long as the diode is really designed as a
stripe, it is justifiable to restrict oneself to a 2D calculation and to simulate
only half of the structure. This is also the case when the n^+ well contact
is designed as a ring around the p^+ anode, provided that the depth of the
structure, which is not depicted here, is much larger than the width of this
2D cross-section.

the process technology itself which represents the limiting factor. The situation
becomes even more difficult when the design of the diode should both be optimized
with respect to ESD, and should also fulfil the very stringent requirements of the
circuit design, when, for example, the rf performance is critical. In such cases,
one has to find a trade-off between ESD robustness and capacitive load. Typical
compromises are:

- The size of the pn junction area (parameter w_{p+} in Figure 5.45) should
  be large for ESD reasons to reduce current density, but small to limit the
  capacitive load.

- In general, a diode can be implemented either as STI or poly-bounded. The
  effective junction width is larger for the poly-bounded diode but the diode
  could become more leaky.

- As the distance L between the electrodes increases (Figure 5.45), the voltage drop across the n well resistance increases and the clamping capabilities
  worsen. However, the shorter the spacing between cathode and anode, the
  higher is the inter-metal capacity (Figure 5.46) between the contacts degrading possibly the rf performance.

- The arrangement of the metal connections has to be such that a good metal
  connectivity guarantees a good heat sinking and low resistive voltage drop.
  However, this measure again increases the inter-metal capacitance.

In principle, all these effects and possible design improvements for the structures can be treated by means of device simulation. In the following, this is
illustrated for a typical diode as depicted in Figure 5.46.

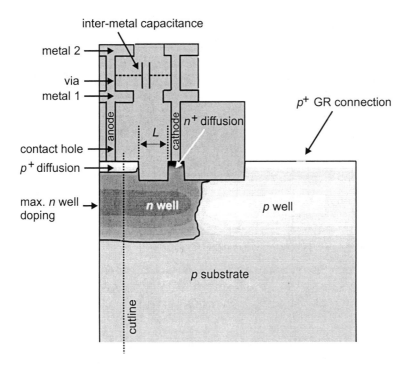

Figure 5.46: Topology and doping profile distribution of an STI-bound p^+ in n well diode used for 2D device simulation (see colour section from page 269).

Variation of the design parameter L

The design parameter L determines the forward bias high current IV characteristic. Under low current injection, the voltage drop across the diode occurs mainly over the pn junction, and the effect of a serial resistance along the n well is negligible. When the high injection region is reached, the additional voltage drop across the n well must be taken into account, leading to a decrease in the slope of the IV characteristic below the expected value of an ideality factor of $n = 1\text{--}2$ (Sze, 1981). Obviously, this series resistance worsens the clamping capability of the diode under ESD.

The degradation of the clamping capability of the diode is investigated for two diodes with a distance L of 1 µm and 2 µm. For the same injection conditions the voltage drop across the device with the larger L is increased by almost 15 % (Figure 5.47). If this is tolerable for the ESD protection concept, i.e. if it is compatible with the ESD design window, then the larger L can be used for an improved rf performance, in case the inter-metal capacitance dominates.

Variation of the STI depth

Inspection of the current distribution inside the device reveals the surprising result, that most of the current flow takes place not along the maximum of the n

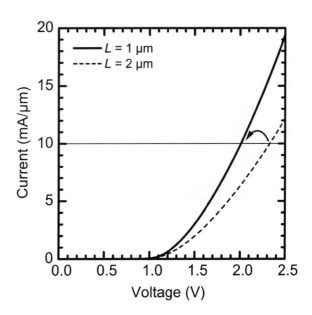

Figure 5.47: *IV* characteristic obtained by device simulation of a "standard" diode ($L = 1$ µm) compared to a diode in which L is doubled.

well doping concentration in the depth, but directly beneath the STI interface (Figure 5.48). This indicates that the on-resistance of the diode can strongly be influenced by the STI depth, even if this is significantly smaller than the depth of the well. This is confirmed by the results depicted in Figure 5.49.

Variation of the n well doping profile

Because of the large voltage drop leading to a high power dissipation at the p^+n junction, the ESD robustness of the diode in reverse direction is rather weak. To avoid any triggering of the breakdown under reverse bias, it is better to increase the breakdown voltage of the p^+n well junction. As the breakdown voltage is rather a process than a design issue, an appropriate doping profile has to be determined in the process technology. (Figure 5.50). A shift of the n well doping peak deeper into the p bulk e.g. by increasing the energy of the n well implant, reduces the absolute amount of doping species at the pn junction and increases the breakdown voltage as shown in Figure 5.50. However, this measure has to be balanced with the latch-up performance of the technology (Bargstaedt, 2001). The trade-off can be found by device simulation.

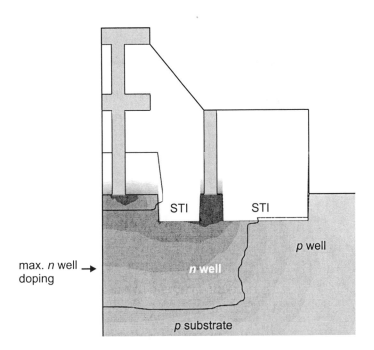

Figure 5.48: Current distribution in an STI bounded diode with $L = 1\,\mu m$ under forward biased high current injection conditions (see colour section from page 269).

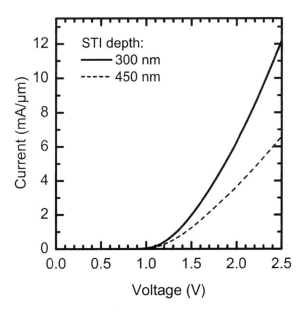

Figure 5.49: IV characteristic of diodes with the STI depth varied. Increasing the depth of the STI pushes the current deeper into the well, making the current path longer and worsening the clamping capabilities of the diode.

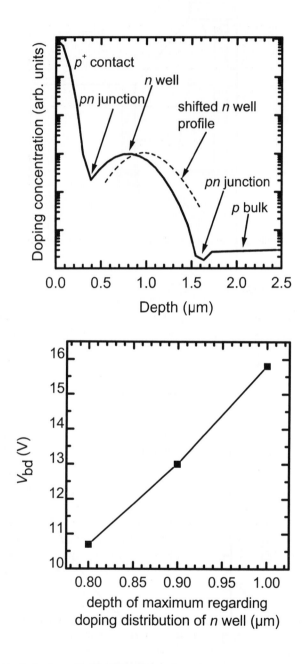

Figure 5.50: Vertical doping distribution from the p^+ contact via the n well to the p substrate (see cutline in Figure 5.46). The distribution of the n well profile has been shifted in an experiment to increase the breakdown voltage.

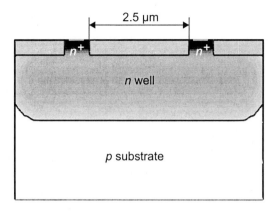

Figure 5.51: Cross section of an n well resistor of 2.5 µm length in a p substrate (see colour section from page 269).

5.5.2 n well resistor

Resistors are widely used in ESD protection concepts. The most prominent examples are de-coupling resistors of first and second protection stage for protecting input circuits, or realizing some ballasting effect for NFET devices. There are several ways to generate a resistor in a silicon process technology using either the

- n well (in a p substrate technology),

- (un-)silicided n^+ or p^+ poly,

- (un-)silicided n^+ or p^+ diffusion.

Physically the most important differences are their sheet resistance values, expressed in ohms per square (Ω/sq), their heat dissipation capabilities and their breakdown behaviour. Using an n well resistor as an example, we will discuss the relevant effects which must be considered for ESD protection development.

The cross section of an n well resistance shows an n^+nn^+ structure, which consists of the n doped n well and two highly n^+ doped contact regions (Figure 5.51). The STI between both contacts prevents the current from flowing close to the silicon surface.

The TLP measured IV characteristic of this device can be divided into three different sections, which are the ohmic or linear regime, the saturation regime and the snapback regime (see Section 2.4.5 and Figure 5.52). In the linear regime, the total resistance of the resistor depends on the sheet resistance of the n well diffusion, which is again a function of the shape of the n well doping profile. In the saturation regime, the linear relationship between electric field and carrier drift velocity is no longer valid. Owing to the velocity saturation of the charge carriers, an increase in the electric field no longer increases the current – at least, not to the same extent as in the linear regime. At higher current densities the

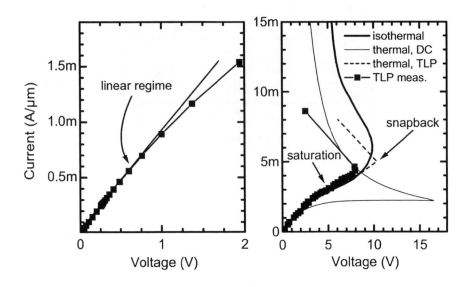

Figure 5.52: *IV* characteristic of an *n* well resistor, showing the ohmic (linear, left), saturation, and snapback regimes (right). The technology's nominal *n* well sheet resistance is 400 Ω/square.

structure exhibits a very large electric field at the n^+–n well junction of the hot pin (Figure 5.53). This reverse biased nn^+ junction has a dramatic impact on the subsequent trace of the *IV* curve, because in its vicinity avalanche multiplication is triggered. The resulting strong growth of the electron and hole density leads to a charge modulation in the region of the *n* well and, consequently, to a non-thermal snapback of the device characteristic. When the resistor is driven into the snapback mode, usually irreversible damage occurs. Device simulation can be used to define appropriate design values for the width and length of the device to ensure that the current density under ESD stress is kept below the critical value for snapback. The observed deviation between simulated and measured values for the snapback voltage in Figure 5.52 are due to the difference between modelled and real doping profile at the nn^+ junction being responsible for the electric breakdown.

In a second experiment, the electro-thermal coupled simulation of the *n* well resistor has been repeated. Instead of applying short TL pulses, the curve has been traced now using DC electrical boundary conditions. This significantly changes the shape of the *IV* characteristic (Figure 5.52).

Inspection of the interior of the device under high current injection conditions reveals that at the position of the reverse biased n^+-to-*n* well junction, the power dissipation causes a significant temperature increase inside the structure. In Figure 5.54 the temperature profiles of the 100 ns TLP pulse and a DC current are compared at a current density of 2 mA/µm. Under short pulse conditions, the

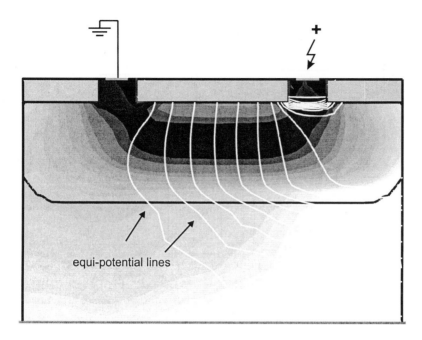

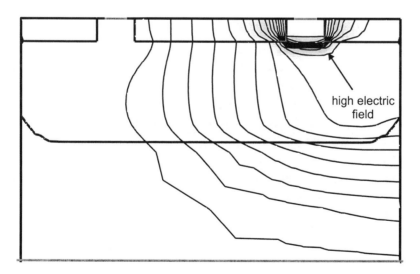

Figure 5.53: Top: Current flow and equipotential lines for the n well resistor under bias conditions. Bottom: Distribution of the electric field inside the structure prior to snapback. As the electric field increases at the n well n^+ contact of the hot pin with increasing voltage, there is an increase in the electron/hole generation, due to avalanche, which finally gives rise to the snapback of the device (see colour section from page 269).

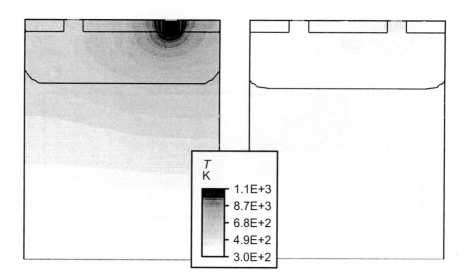

Figure 5.54: Temperature inside the resistor under DC (left) and transient (right) conditions for 2 mA/μm (see colour section from page 269).

heat flow from the hot spot into the drift region of the resistor body is small, while under static conditions the complete bulk can be heated up. Since the carrier mobility degrades with increasing temperature, the IV curve shows a higher differential resistance in the linear and saturation regime with increasing pulse width.

As the temperature increases with rising current density across the n^+-to-n well junction, a region of thermally generated carriers spreads out across the drift region and a part of the resistor device may finally become intrinsic (Figure 5.55). At this point a strong current contribution appears, owing to thermally generated electrons and holes leading to a thermally induced snapback of the device. As shown in (Figure 5.55), the contribution of avalanche generated carriers to the number of totally generated carriers is significantly smaller. To conclude, under DC stress conditions a destructive snapback might be triggered at lower current densities than under transient conditions (Figure 5.52). This emphasizes the correct treatment of the pulse. However, the avalanche triggered snapback always provides the upper limit of the trigger current.

The general trend is that resistors with low specific resistance like n^+/p^+ diffusion resistors tend to break down thermally because of the higher current densities at equivalent electric fields. Resistors with high sheet resistance like n well resistors favour an electrical breakdown as long as short pulse conditions prevent heat propagation into the bulk of the device. As Figure 5.52 shows, in this case it is justified to apply an isothermal simulation. An exception are resistors with low breakdown voltage of the diffusion-to-bulk junction where a breakdown of the junction at the hot pin might occur.

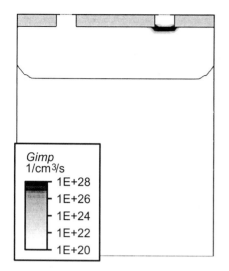 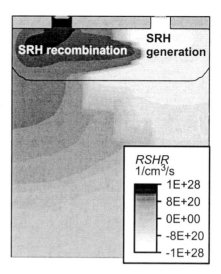

Figure 5.55: Impact ionization (left) and (SRH)-thermal generation(right) at snapback
under DC conditions for 2 mA/μm. Here the thermally induced carriers
cause snapback (see colour section from page 269).

5.5.3   BJTs

Bipolar junction transistors (BJTs) are omnipresent in the ESD concept of all possible technologies. While the parasitic *npn* device associated with CMOS transistors is the workhorse of many ESD protection concepts in CMOS technologies, the
native *npn* transistor is the counterpart for modern bipolar or BiCMOS technologies. Because of its breakdown and snapback characteristic, it provides excellent
voltage clamping, and is in many cases better controlled and more stable against
technology drifts than, say, thyristor structures. The behaviour of the parasitic
bipolar transistor formed in FETs of CMOS technologies is extensively discussed
in Section 5.4. In many respects, lateral BJTs in bipolar technologies behave quite
similarly. Therefore, in the current section an effect is discussed which is typical
for BJTs in bipolar or BiCMOS technologies (Grutchfield, 1966; Streibl, 2002).

Many modern bipolar and BiCMOS technologies offer their *npn* transistors in
two flavours: high voltage (HV) transistors which are targeted at a high I/O signal
range rather than an optimized rf performance, and rf *npn* devices which usually
sacrifice operational voltage range (i.e. open base breakdown voltage V_{CE0}) to
gain as high a f_T as possible (Voldman, 2001). In the latter case, one or more
additional selective implants through the emitter window, usually called selectively
implanted collectors (SIC), are widely used to block base pushout and to achieve a
low-ohmic collector. For clarification, Figure 5.56 shows schematic cross-sections
of typical rf and HV vertical *npn* devices. A characteristic feature of these devices
is the so-called base-pushout, or Kirk effect, which leads to a complex high current
behaviour. After the triggering of the bipolar action by sufficiently high avalanche
currents, and the associated snapback to the holding voltage V_h, the transistor

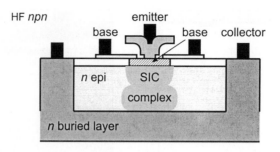

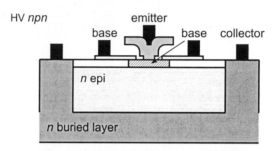

Figure 5.56: Cross sections of vertical rf (top) and HV (bottom) *npn* devices. The selectively implanted collectors (SIC) consist of a medium level *n* doping which reduces the collector resistance and shifts the onset of the base pushout to higher current densities.

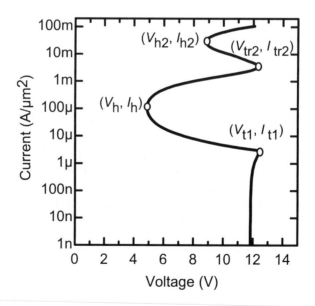

Figure 5.57: *IV* simulation of rf *npn* with double snapback due to base pushout. The current density shown refers to the emitter window area.

can snapback a second time at I_{tr2} and reaches again a non-destructive, reversible holding point(I_{h2}, V_{h2}), because of the base pushout (Figure 5.57).

2D device simulation can reveal the mechanisms at work at critical points of the IV trace, such as the field distribution in the base-pushout regime or the transition of the base-pushout phase to the real low-ohmic breakdown mode via a non-thermal "second snapback".

At high collector current densities the collector–base (C–B) depletion zone of an npn bipolar junction transistor is flooded with electrons. The base-pushout occurs when the collector current electron density exceeds the background doping level N_C on the collector side in the C–B depletion region. Under such conditions, the background n doping is effectively screened by the free carriers, and the field free base region is stretched towards the higher doped collector sections, thus pushing the depletion region out into the collector and leading to an effective increase of the base width. Along with the depletion zone, the maximum electric field at the C–B boundary is shifted into the collector (Figure 5.58). The critical current density I_C for the onset of base pushout can be written as:

$$I_C = qAN_C v_{\text{sat}} \qquad (5.20)$$

where A is the area of the base, N_C the background doping of the n^- collector region and v_{sat} the saturation velocity of the electrons in this region.

If the voltage drop is now sufficiently high across the lowly doped n^- collector region, i.e. higher than V_{tr2}, the transition to a transistor with high base width but much lower collector resistance will be accompanied by a second, non-thermal snapback, which is as reversible as the first snapback.

The SIC implants shifts the onset of the Kirk effect to higher current densities, since the critical collector current I_C is determined by the dopant concentration N_C in the collector depletion zone. This is done to provide the rf circuit designer a wider operational current range. In the case of the HV npn, with a very lightly n doped epitaxial layer (n-epi) touching the p base, the Kirk effect plays a dominant role even at very small current densities.

Once the base pushout is fully established, i.e. once the npn has gone through the base pushout snapback, there is virtually no difference between devices with different SIC/n-epi structures, provided that they have the same base and n buried layer complex. Even the electric field distributions will look alike, so that the rf and HV npn transistors high current behaviour will match. This means that despite the lower V_{CE0} voltage of the rf transistor, the voltage drop over the device at higher ESD current densities (i.e. beyond base pushout) is comparable to the corresponding voltage of the HV npn (Figure 5.59). Particularly for rf ESD protection concepts, this is a most useful result. In many cases it is favourable to use HV npn protection elements, because they have larger V_{CE0}, and can therefore be used for a higher signal or supply voltage range, while still providing the same voltage clamping for ESD.

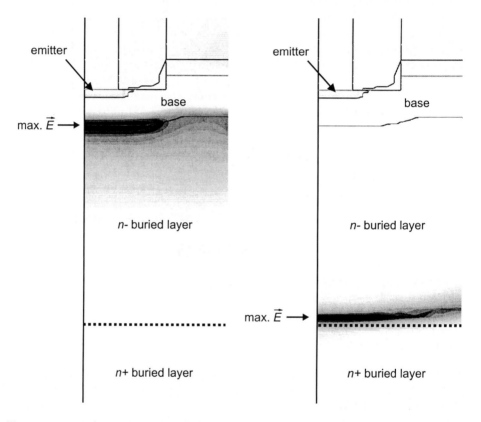

Figure 5.58: Simulation of the electric field distribution E before (left) and after (right) base pushout. The distribution of the maximum electric field moves from the pn junction of the base–collector junction to higher doped collector sections. A selectively implanted collector (SIC) has been used (see also Figure 5.56 and colour section from page 269).

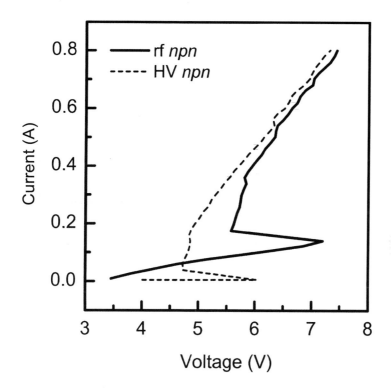

Figure 5.59: Matching of rf and HV *npn* transistors at high current levels measured for a
0.25 μm BiCMOS technology.

5.5.4   SCRs

As well as for reasons of area optimization and capacitive load, SCRs are increasingly used to keep pace with an ESD design window that shrinks steadily with
every new CMOS technology.

As for the NFET, the important parameters for the purposes of ESD development are trigger voltage and trigger current, sustaining point, clamping behaviour,
and ESD robustness per area. Specifically for SCRs, another question related to
ESD performance is its intrinsic response time and clamping capability during
very fast ESD events like CDM.

There are several ways to overcome the major drawbacks of a standard SCR,
namely the excessive trigger current and trigger voltage, which might make the
device incompatible with the constraints of the ESD design window of a particular
technology. By integrating internal (Amerasekera, 2002) or external (Russ, 2001)
trigger elements into the SCR, triggering of the SCR can be achieved at much
lower current and voltage values. The necessary charge carrier modulation of
substrate and *n* well regions can be achieved by charge injection as a result of the
breakdown of the trigger device. Another critical parameter with respect to the

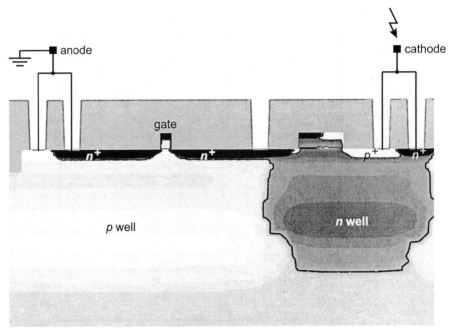

Figure 5.60: Cross-section showing the doping profile of a LVTSCR (see colour section from page 269).

SCR is the sustaining point, which can be lower than the operational voltage of the chip, eventually causing a latch-up problem. Based on a low voltage triggered SCR (LVTSCR; see Figure 5.60), some device-internal processes under ESD stress conditions are highlighted below. This type of SCR consists of a *npn* transistor determined by the lateral arrangement of the layers n well/n^+ drain (collector), p well (base) and n^+ source (emitter), while the *pnp* is formed by the p^+ diffusion of the cathode (emitter) n well (base) and p well (collector). By integrating an NFET device into the SCR, the breakdown voltage of the structure is no longer determined by the very high breakdown voltage of the n well to p well junction but by the breakdown voltage of the n^+ diffusion (drain) to p well junction of the NFET device.

By comparing a LVTSCR and an NFET one sees that both devices trigger at about 6 V, but the NFET device shows a much higher differential resistance compared with the SCR (Figure 5.61). The LVTSCR therefore shows the better voltage clamping. The SCR operation is not fully established with the triggering of the NFET device but at some higher current levels. The reason for this is that, owing to the set-up of the LVTSCR, the current of the *npn* structure is not efficiently injected into the *pnp* structure, making it more difficult to trigger the *pnpn*. The sustaining point of the device is about 2 V, which is sufficient, with respect to the products operating voltage (1.5 V+ 10 %), to avoid unintentional latch-up of the LVTSCR.

Considering the electron and hole current density distributions when the SCR

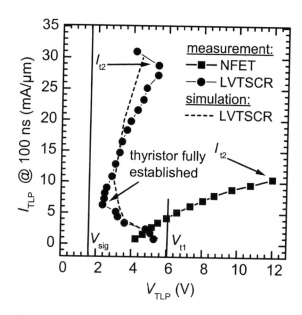

Figure 5.61: High current IV characteristics of an NFET and an LVTSCR (width = 100 μm). The simulated and measured characteristics of the LVTSCR do not match as perfectly as in the case of the NFET device due to the critical calibration of the carrier lifetime parameter for SCRs.

has fully triggered, the operation of both the npn and the pnp transistor can clearly be seen (Figure 5.62). While the npn is characterized by a large electron current flowing at the surface of the device, the hole current deeper in the bulk of the structure indicates the activity of the pnp. By studying the hole current distribution at the n well/p well junction, the device simulation can be used to optimize the trigger behaviour. As the transition n well to p well below the n^+ diffusion is not specifically monitored, process fluctuations may lead to an uncontrolled shift in the ESD parameters of the SCR. In this situation, TCAD tools can help to find a robust design which is less sensitive to the process variations.

The simulated and measured current densities at the device destruction level are much higher for the LVTSCR than for the single NFET device, since heat dissipation is much lower and the affected volume, in which the power is dissipated, is larger (Figure 5.63).

Because of the large volume of the p well and n well of the thyristor which needs to be flooded by carriers before the element can be triggered, the intrinsic response time of an SCR to an ESD event is larger compared to an NFET device. Particularly for very short ESD events, the turn-on speed of an LVTSCR device is a critical topic and must be tested. An NFET and an LVTSCR, both having the same width of 100 μm are exposed to a 100 V CDM pulse with 0.4 ns rise time, as discussed in more detail in Section 5.6.2. Looking at the response of the devices in terms of the voltage waveform, one can see that the triggering of the

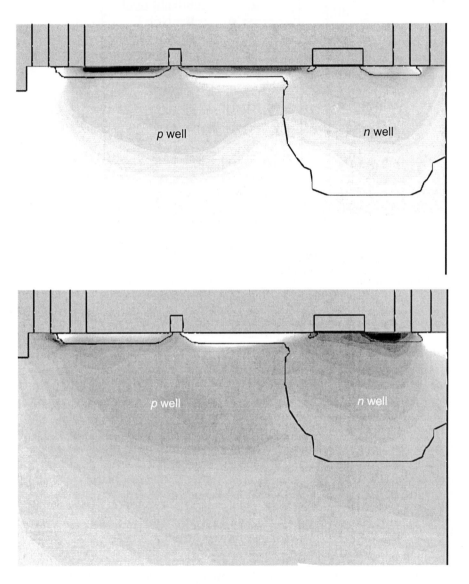

Figure 5.62: Distribution of electron (top) and hole (bottom) current density inside the LVTSCR under ESD stress conditions indicating *npn* and *pnp* bipolar operation of the device (see colour section from page 269).

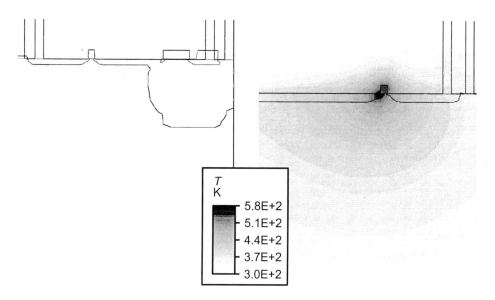

Figure 5.63: Temperature distribution of an LVTSCR (left) and NFET (right) under same ESD injection conditions (see colour section from page 269).

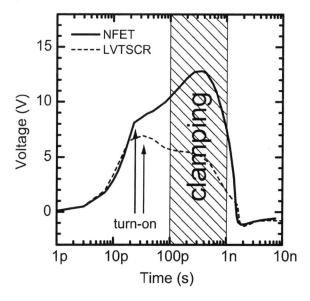

Figure 5.64: Voltage waveform of NFET and SCR devices as a result of the CDM current pulse with a pre-charge voltage $V_{CDM} = 100$ V, (see also Section 5.6.2). Turn-on time and clamping behaviour are the relevant parameters describing the ESD performance with respect to very short pulses. The turn-on was derived from an analysis of the transient current distributions in the cross sections of the devices.

NFET device occurs already after 10 ps (Figure 5.64). The SCR triggers about 10 ps later, but it is still fast enough to clamp the voltage sufficiently during the ESD pulse. In CDM, current levels of several amps occur, and therefore the much lower differential resistance of the LVTSCR in comparison to the NFET helps to clamp the voltage at much lower values for comparable device size.

5.5.5   Self-protection in sub-100-nm SOI technologies

In the examples presented in the preceding sections, ESD device simulation was used to optimize the ESD robustness and clamping capability of dedicated protection devices. However, ESD device simulation can be used as a powerful tool to optimize technologies with respect to ESD, without the need of processing costly hardware split lots. Examples how process parameters influence the ESD behaviour have already been discussed, see for example Section 5.5.1. In fact, even if applied to a completely new class of processes, the ESD engineer can benefit from the ESD device simulation as it yields fundamental properties and at least qualitative understanding. An example of this particular application of ESD device simulation is discussed in the following.

Recently, several world records in transistor scaling were published pointing the direction for the CMOS technology in the second half of this decade (Yu, 2001). Even if the issues of leakage and performance can be reconciled by new approaches, for example by the use of SOI (silicon-on-insulator) technologies, the non-scalable requirements of ESD hardness have to be considered to avoid a road block for the down-scaling of the CMOS technologies. SOI technologies seem to be particular critical for ESD as the reduced thermal conductivity of the SiO_2 which surrounds the active area (see Figure 5.65). To proof this, a semi-quantitative analysis is performed by comparing the maximum temperature during an ESD pulse and "standard" bulk silicon with a fully-depleted SOI (FD-SOI) and an SOI process variation with raised drain/source diffusions in a 0.1 µm technology. As a vehicle for the study a standard NFET with minimum gate length and unsilicided diffusions is used.

The electric field, current density and temperature distribution during the high current transients of a TLP pulse are extracted from a 2D ESD device simulation. All process and transport models have been calibrated for a 0.1 µm CMOS technology. The SOI devices with and without raised S/D are deduced from this bulk technology by the introduction of a buried oxide layer and an extension of the source and drain region in the device simulation input file. Although this approach does not give an exact electronic copy of a real existing device, the main features of the process variations are reproduced sufficiently.

Compared to the standard bulk process, the dominant consequence of FD-SOI in modern technologies is that the large ESD current is forced through an only few 10 nm thick silicon layer, see Figure 5.66. That silicon layer is thermally well isolated by the surrounding oxide (Figure 5.65). Therefore, it is not astonishing that the maximum temperature of a silicide blocked NFET rises to 820 K compared to 420 K in bulk material at a current density of 3 mA/µm (Figure 5.66). As a consequence, the failure threshold drops from 9 mA/µm of the bulk device to

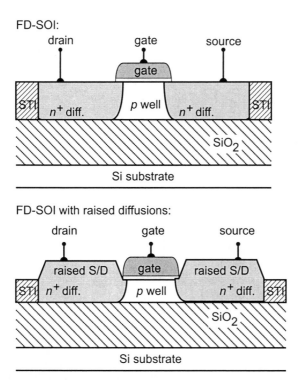

Figure 5.65: Schematic cross-section of an NFET in an SOI technology. Top: Fully depleted (FD) SOI; bottom: FD-SOI featuring raised diffusions.

3 mA/µm. In other words, provided a homogeneous current flow along the width of the device, the device width of the NFET (and the resulting area, too) must be three times larger in order to achieve the same ESD robustness as in the standard CMOS process. The problem of the current concentration in the very thin active Si layer appears to be a problem for all other possible protection elements, too.

Silicon epitaxy on the source and drain region leads to a thicker conductive layer and an effectively lower sheet resistance of the drain and source electrodes. At the same time the heat sink is improved due to the increased amount of silicon attached to the region of maximum temperature. Both leads to a reduction of the maximum temperature compared to the FD-SOI. The maximum temperature of a silicide blocked NFET at 3 mA/µm is decreased to 640 K (Figure 5.67) and the failure threshold moves up to 4.5 mA/µm, 50 % higher that the value for the FD-SOI process. However, compared to the standard CMOS process the ESD robustness is decreased by a factor of 2, leading to a at least two times larger area consumption of the ESD protection elements. Although the ESD robustness of SOI process variation decreases compared to the corresponding bulk process, from this basic study there seems to be no fundamental blocking point for an ESD concept in SOI technologies. This is well in line with the conclusions found in literature (Duvvury, 1996).

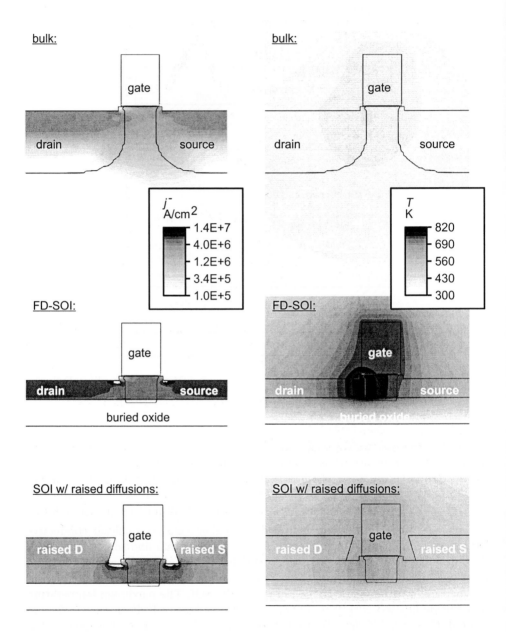

Figure 5.66: Current density (left) and temperature (right) during an HBM pulse forcing
3 mA/µm into the device after 100 ns in process variations with feature size
$l_{\mathrm{g}} = 0.1$ µm, $d_{\mathrm{Si}} = 50$ nm. Top: (Standard) bulk CMOS process, middle:
FD-SOI, bottom: FD-SOI featuring raised diffusions. The highest current
distribution and the highest temperatures occur in the FD-SOI.

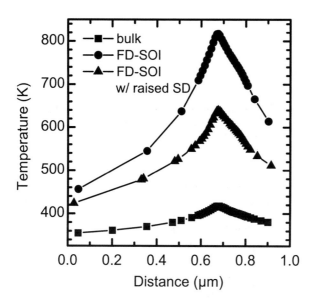

Figure 5.67: Lateral temperature profile during an TLP pulse forcing 3 mA/µm into the device after 100 ns in process variations with $l_g = 0.1$ µm along a cutline through the maximum temperature parallel to the gate.

5.6   Simulation of different ESD stress pulses

In this chapter, we have hitherto restricted ourselves to considering only short square current pulses as discharge waveforms. The intention has been to characterize the ESD protection elements in the high current regime to derive essential electro-thermal parameters like trigger voltage and sustaining point in an ESD typical time domain. Among the set of electro-thermal parameters mentioned above is the current-to-failure I_{t2}, which represents the maximum current that can be handled by the structure for a certain pulse duration. There are many reasons for a finite I_{t2} and the subsequent breakdown of the structure, ranging from thermal overload of a *pn* junction to the electrical breakdown of the gate oxide. One can imagine that the wide range of possible ESD discharge waveforms will lead to different types of failure mechanisms, leading to totally different results. Therefore the focus of this section is to provide insight into the device response when it is exposed to two of the most important discharge events, namely the Human Body Model (HBM) and the Charged Device Model (CDM).

5.6.1   HBM

As explained in detail in Section 1.2.2, the Human Body Model represents a discharge of a 100 pF capacitance through a 1.5 kΩ resistor. The resulting waveform of a distinct HBM tester can be modelled using the lumped-element circuit introduced in Figure 1.3.

The current and voltage response of the reference ggNFET transistor in a
0.35 µm technology for a pre-charge of $V_{HBM} = 3$ kV is depicted in Figure 5.68.
Within 2 ns, the parasitic transistor has triggered. After approximately 20 ns,
which roughly represents the rise time of the pulse in this particular tester, the
current has reached its maximum level of nearly 1.8 A. The current then starts
to decrease, until the HBM capacitance has nearly discharged. In this regime,
the voltage initially decreases until the sustaining point V_h has been reached at
a low current level. The voltage at the drain node of the DUT then increases
again, since the ggNFET switches back from the parasitic bipolar to the diode
mode. The remaining accumulated charge at the drain node finally decays via the
leakage current of the ggNFET.

Parallel to the voltage and current waveform, the maximum temperature inside
the device is plotted. The maximum temperature of the device does not strictly
follow the power dissipation. While the maximum power is dissipated 20 ns after
the beginning of the pulse, the highest temperature is reached just after 100 ns.
After 20 ns, the device is still being heated, even though with a reduced power.
This is a result of the non-equilibrium conditions during the pulse.

For later analysis of device breakdown it is useful to combine the transient
voltage and current waveform to create a transient IV characteristic for the ggN-
FET and indicate, at which points maximum power dissipation occurs and when
the maximum temperature inside the device is observed (Figure 5.69). Compared
to the quasi-static IV characteristic obtained by TLP, the resulting curve shows
a hysteresis for several reasons:

1. The temperature inside the structure continuously changes and therefore all
   other device parameters are under continuous change.

2. Trigger processes and behaviour are time dependent, being influenced by
   displacement currents.

To determine the HBM robustness of the devices by means of simulation, the
TLP testing criterion is adapted to this situation:

$$I_{HBM,sim} = \min \quad \{I@(T_{max} > T_{crit}), \qquad\qquad (5.21)$$
$$I@(\text{locally; occurrence of 2nd breakdown})$$
$$I@(E_{field,GOX} > E_{crit})\}$$

Monitoring the temperature as proposed in Equation 5.21 is a simple task,
but the detection of 2nd breakdown is rather difficult when looking at a typical
transient IV characteristic like in Figure 5.69.

The analysis becomes easier in the case of a TLP stress. As discussed in
chapter 5.4 the second breakdown can be clearly detected in the waveform of a TL
pulse. A sudden drop of the voltage is seen in the – after the initial overshoot–
steadily increasing part of the $V(t)$ waveform (Figure 5.70). This also leads to a
change of the slope in the extracted IV characteristic, if the drop occurs within the

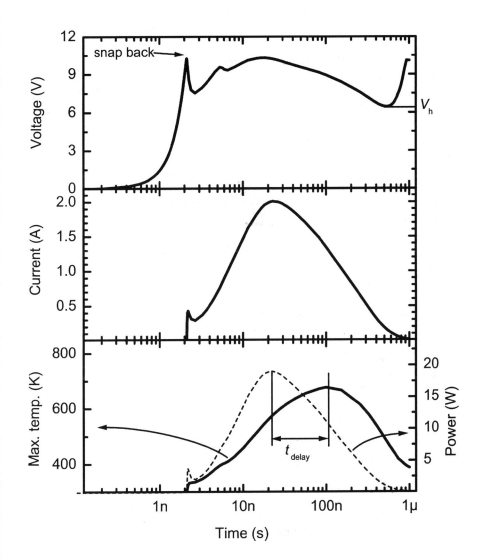

Figure 5.68: Simulated voltage (top), current (middle) and maximum temperature/power (bottom) for the reference ggNFET in a 0.35 µm technology for a V_{HBM} = 3 kV pre-charged capacitance C_{HBM}. The maximum temperature does not coincide in time with the maximum power dissipation but is delayed by t_{delay}.

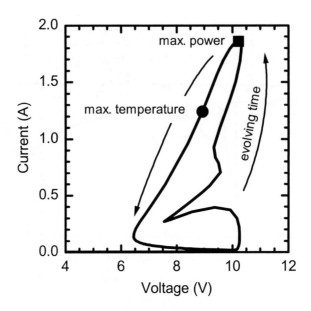

Figure 5.69: Transient IV characteristic of a ggNFET resulting from a pre-charged C_{HBM} of 3 kV. The point along the trace where maximum power dissipated and maximum temperature is observed, are indicated.

analysis window of the voltage extraction (see also Chapter 1.4.1 and Figure 5.35). However, for the complex waveform of the HBM pulse the detection of the second breakdown is more difficult and such a straight forward criterion is not applicable. An alternative approach is borrowed from the theory of phase changes. As shown in Esmark (2001), the analysis of a Poincaré map (Perko, 1991) is a suitable tool for detecting a second breakdown as a consequence of a device instability. From a more general point of view, stability of a dynamic system means that exposing the system to a slightly different external environment should cause only small changes in the intrinsic parameters of the system. Large changes of the internal parameters indicates a phase transition or in our case a second breakdown.

A distinct point of the trajectory of the pules like the point for the occurrence of maximum temperature or maximum power dissipation can be extracted from the simulation (Figure 5.69) and depicted in a IV diagram with increasing stress level. The resulting graph (the so-called hyperplane Σ (Perko, 1991)) gives an information about the state of the investigated device. When the second breakdown occurs the point e.g. at the occurrence of maximum temperature suddenly shifts along the trajectory and its IV characteristic depicts a kink (Figure 5.71). Of course, this approach can also be used used in the analysis of a TLP waveform, where e.g the point of the maximum power dissipation jumps from the end of the pulse to point in time correlated with the second breakdown (Figure 5.70). If a sufficient number of stress levels are used as sampling points of the IV characteristic an HBM threshold voltage can be extracted with this procedure. This is shown

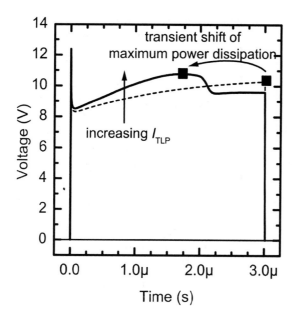

Figure 5.70: Voltage waveforms for the ggNFET transistor resulting from increasing TL pulses. The transiently dissipated power, in this particular case correlating to the voltage waveform, reaches a maximum at the end of the pulse. The reason is an internal heating effect, which influences internal device parameters like carrier mobility and avalanche generation. If the device shows a second breakdown, the maximum power is no longer dissipated at the end of the pulse, but somewhat earlier. In this case, the parameter "maximum dissipated power" has experienced a transient shift.

for a ggNFET in a 0.18 µm technology which HBM robustness is limited by second breakdown. The kink in the hyperplane Σ representing the path of one intrinsic parameter of the maximum temperature, occurs for a pre-charge $V_{HBM} \approx 2$ kV (Figure 5.71).

Another application of equation 5.21 to detect device breakdown under HBM stress conditions is highlighted for the example of a ggNFET in a 0.35 µm CMOS technology. From the investigation of the evolution of the maximum temperature in Figure 5.72, it can be verified that for this particular device the decisive failure criterion in Equation 5.21 is the exceeding of the melting point of silicon rather than a second breakdown.

The HBM robustness is estimated to be between 5.1 kV and 5.2 kV from the device simulation, because at this level the temperature inside the device exceeds the melting point of silicon. Experimentally, a HBM robustness of 5.8 kV was observed, which is about 10 % higher than in the simulation. This is a good agreement and proves that the simulation criterion gives indeed a worst case estimation. The criterion regarding detection of second breakdown is not met for any pre-charge value V_{HBM} of the particular NFET device below 5.1 kV.

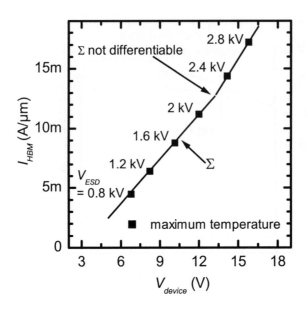

Figure 5.71: Hyperplane Σ of the maximum temperature $T_{\max}$ for the ggNFET transistor in a $0.18\ \mu m$ process technology. A kink in the characteristics can be observed between $V_{\mathrm{HBM}} = 2\ kV$ and $2.4\ kV$, which is identified as second breakdown.

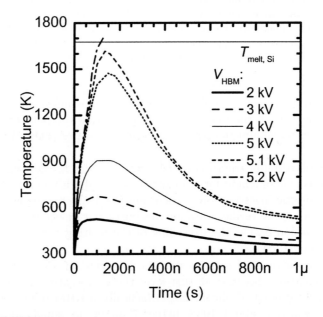

Figure 5.72: Transient temperature evolution for different HBM pre-charges V_{HBM}. For $V_{\mathrm{HBM}} > 5.1\ kV$, the maximum internal temperature exceeds the melting point of silicon which determines the HBM robustness (ggNFET, $0.35\ \mu m$).

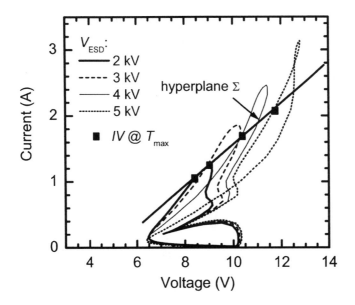

Figure 5.73: The points indicating maximum temperature for differently pre-charge values V_{HBM} are used to generate a hyperplane Σ (ggNFET, 0.35 µm).

The analysis on base of the Poincaré map results in a hyperplane which is nearly a straight line without any (non-differentiable) kink (Figure 5.73). This is an indication that the intrinsic parameter always appears at the same point along the trajectory of the device regardless of the pre-charge V_{HBM} and therefore no phase transition in the form of a 2nd breakdown took place.

5.6.2 CDM

Modelling the CDM discharge current waveform

In contrast to the human body model, the CDM is a very fast discharge within a few nanoseconds, which leads to significantly higher current levels. This different behaviour therefore causes different failure mechanisms. The predominant failure in CDM is dielectric breakdown caused by the possibly high voltage drops across devices or parasitics in the IC. Thermal damage is in general less important.

The CDM discharge waveform mainly depends on the device package, the pin position, the wire bonding and the testing environment (see Section 1.2.4). Figure 5.74 shows an experimentally determined 500 V CDM discharge which has been recorded for one of the pins of a simple C-DIP-40 package according to the field-induced method. As can be seen from the observed discharge curve, the rise-time of the pulse is less than 200 ps while the pulse duration is no longer than several nanoseconds. Further on, an undershoot can be observed, again different to HBM. To model the resulting discharge waveform in a device simulator, the

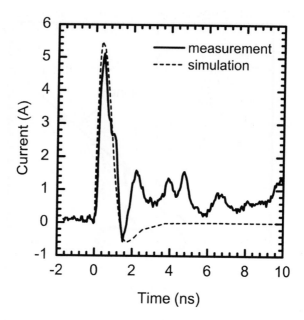

Figure 5.74: The measured pulse shape of a CDM discharge of 500 V for a certain pin of a C-DIP-40 chip. For modelling purposes, the pulse waveform has been generated using a LEM with fitted parameters for the different parasitics in the test setup. In the distinct package, reflections occur which cannot be reproduced by the simple LEM from Section 1.2. For a meaningful simulation, the maximum positive and the maximum negative current during the CDM event must be modelled correctly, therefore, the parameters of the LEM are tuned to fit the first positive and the first negative current peak. The experimentally observed peaks after the first 2 ns are caused by reflections of the package/tester system and are not considered further.

curve can be generated using the equivalent circuit discussed in Section 1.2.

Targets of CDM Device Simulation

Before starting CDM device simulation, we should briefly define the targets for this kind of investigation, since they have consequences for the simulation approach. As already addressed in the introduction of this section, CDM causes a different kind of failure mode than HBM. While HBM addresses failure modes related to thermal overstress of a certain region of a device, the CDM stress predominately causes electrical breakdown of oxides as a result of overvoltage stress. Because of the short pulse duration, the amount of dissipated energy is rather low and only minimal heating occurs. Comparing the simulation of the IV characteristic (disregarding electro-thermal effects) with the measured characteristic of very fast TLP operating in the same time regime as CDM, a good agreement can be found, which justifies the above mentioned assumption. The problem with respect to

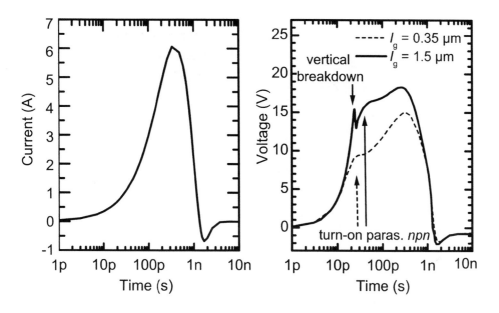

Figure 5.75: Discharging current and voltage waveform of two ggNFET devices of a 0.35 µm CMOS technology with different gate lengths, exposed to a CDM stress of 500 V. The current waveform is taken from Figure 5.74.

a CDM discharge event is that the ESD protection device might be too slow to trigger snapback and clamp the voltage at sufficiently low levels. The term "too slow" means that the risetime of the pulse is much shorter than the devices response time to trigger inherent elements. For example, for an NFET the response-relevant parameter is the base transit time which depends in first order on the gate length of the device. In consequence, for NFET devices having a larger gate length, even the displacement currents of the drain–bulk junction might not allow the collector–base junction to become forward biased and to switch on the parasitic bipolar transistor. With a current amplitude of several Amps, the result will be a large overshoot in the voltage signal, sometimes significantly higher than the static breakdown voltage of the pn junction. Therefore, if a device fails as a result of a specific CDM event, the analysis must be carried out based on a comparison between the observed and maximum tolerable electrical field strengths.

Two ggNFET devices taken from a 0.35 µm CMOS technology, with different gate length (0.35 µm, 1.5 µm), but identical in all other design parameters, were exposed to the fast CDM stress. The shape of the current discharge pulse is independent of the device, but the response of the device in form of the voltage waveform is different (Figure 5.75). The voltage waveform of the short device exhibits a plateau at 20 ps on the rising edge of the pulse.

From investigations of the voltage distributions and current contributions in the device, it can be shown that this plateau arises from the triggering of the bipolar transistor (Figure 5.76). Previously the voltage drop occurs only at the

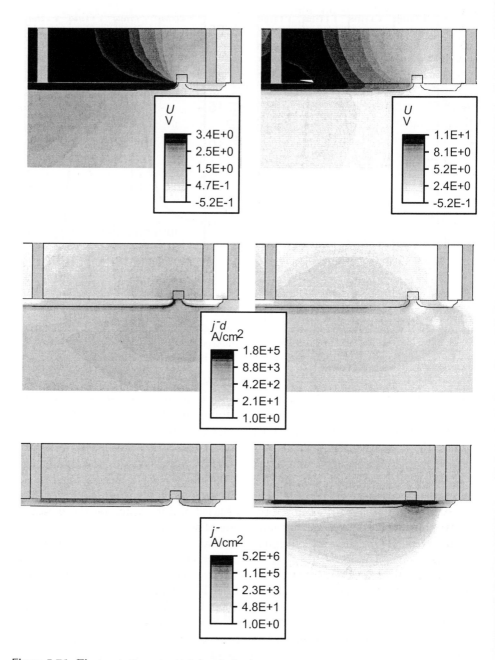

Figure 5.76: Electrostatic potential (top) displacement current density (middle) and total current density (bottom) distribution of the ggNFET with gate length
0.35 µm before, $t = 10$ ps, (left) and after, $t = 50$ ps, (right) triggering of
the parasitic bipolar transistor for a CDM stress of 500 V (see colour section
from page 269).

reverse biased drain to bulk diode and the displacement current represents the main contribution to the overall current density. After the triggering, one can observe the voltage drop across the drain diffusion while the electron and hole transport current take over the dominant role in the total current density while the displacement current becomes negligible small.

For the longer device the plateau is shifted to longer times (40 ps) indicating a later triggering. In addition, a small peak appears at 10 ps which is found to belong to transient redistribution of the displacement currents as a result of the increased gate length (vertical breakdown of the *pn* diode).

In general the voltage overshoot at the peak of the pulse is very high as result of the large discharge current (> 6 A). It can be seen that the clamping capability is drastically worse for the larger gate length. A short gate length is thus clearly preferable for ensuring good CDM protection.

During the very high voltage overshoot at the device nodes, a high electric field strength is generated inside the gate oxide. Although present only for a very short time, the electric field strength might exceed the tolerable limits for a gate oxide. For NFET devices, the experimentally determined upper limit for the electric field for gate oxide stress is about 3.4×10^7 V/cm for a CDM typical pulse length of 1–2 ns (Stadler, 2003). A degradation, or even a breakdown, of the gate oxide is very likely to occur for CDM events, which cause higher electric field strength in the oxides of the devices (Figure 5.77).

Today, we are entering the sub-100 µm region for CMOS technologies, leading to an even further reduction in the oxide thickness. From the comparison in Figure 5.77 one can see that the devices with lower gate oxide thickness show an increased susceptibility to electrical breakdown of the gate oxide, since the derived electric field strengths are higher in comparison to older technology nodes for the same CDM stress conditions.

5.7   Limitations of ESD device simulation

The examples of Chapter 5 have shown that ESD device simulation can give answers to a lot of different questions related to high current injection and ESD stress in semiconductor devices. The need to incorporate 3D device simulation whenever snapback devices are investigated has clearly been demonstrated. But, owing to the enormous computation effort required, 3D device simulation cannot be used as a standard tool in today's engineering flow. As demonstrated, although 2D device simulation lacks the necessary accuracy in the snapback regime, in many cases it reflects a kind of worst-case scenario in these situations, which often is sufficient for engineering. Furthermore, current 3D device simulation also represents only a kind of best-case estimation for ESD device behaviour. It is able to predict current filamentation and device destruction as a result of insufficient ballasting resistance, which is absolutely helpful for deriving a pre-Si ESD protection concept. However, statistical fluctuations and variations of doping distribution and device topology along the device width are not covered in the doping profile, which can lead to device breakdown sooner than expected.

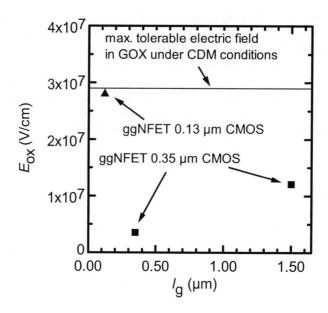

Figure 5.77: Simulated electric field strength of several NFET devices (two with 0.35 μm
and one with 0.13 μm CMOS process technology) during a CDM stress of
500 V. For each technology node, the shorter the gate length the lower the
electric field stress in the oxide. Comparison between different technologies
having different effective gate oxide thickness (here e.g., 0.35 μm: 8.7 nm,
0.13 μm: 2.2 nm) shows that with each new technology generation the problem of gate oxide degradation is becoming more urgent, since the electric
field during a CDM stress is approaching the maximum tolerable electric
field strength of the gate oxide.

As reported in Esmark (2001), there are small discrepancies between simulated
and measured electro-thermal parameters obtained for ESD protection devices under high current injection. In particular, the predicted absolute values of the temperature profile, which are used as a criterion for a device failure, do not perfectly
match the measured data. However, the quality of the calculated results is good
enough to determine tendencies in the device behaviour arising from changes in
the design or process technology of a semiconductor device. This is sufficient for
the development of effective ESD concepts and for device optimisation.

There are also some limitations concerning the selection of devices that can be
simulated with a high quantitative accuracy. The existing models implemented
in the device simulator do not yet properly reproduce devices such as SCRs, with
complex generation and recombination processes happening at different locations
in the device. This is certainly not a fundamental problem of the simulation,
but requires more refinement of the models and their implementation into the
simulator.

Last, but not least, is another problem which is not only associated with ESD device simulation but with device modelling in general. This is the general shrink in dimensions as the CMOS community enters the sub-100 nm range. The simulation approach based on a drift diffusion model might no longer be justified. As Monte Carlo simulations have shown, the influence of ballistic effects on the device characteristic will become more important. Even if the normal operation of, say, FETs can be well reproduced down to a gate length of 40 nm by a drift diffusion approach using an appropriate parameter fitting (Bufler, 2003), the effect on the high current parameters like avalanche multiplication must be carefully investigated.

Summary

The most important findings of this chapter can be summarized by:

- The simulation of the high current densities under ESD stress requires correct treatment of the electro-thermal transport and the heat flow. A consistent approach is given by the reduced electro-thermal transport model. However, this does not include the ballistic effects of the charge carrier.

- Correct use of the device simulator for ESD simulation requires the process-technology-specific calibration of the SRH generation and recombination parameters. Most of the other parameters are usually close to the default values of the TCAD tool. Larger deviations often indicate a wrong doping profile.

- For simulation of a realistic temperature distribution in the device, the correct treatment of thermal boundaries is highly relevant. The bulk of the semiconductor, the covering dielectric layer and the contact holes should be included into the thermal simulation. Beyond that, thermal lumped elements like thermal resistors can be used.

- It has been shown that the high-current IV characteristic can properly be reproduced by device simulation. Depending on the current regime, 2D simulation may be enough, or else 3D simulation must be applied because of current instabilities in the snapback regimes. Owing to the long runtimes, the use of 3D simulation is still very restricted.

- In device simulations there is no simple failure criterion like an increased leakage in a measurement. To evaluate this topic, an approach is given which is based on the most critical value of the maximum temperature, occurrence of second breakdown and dielectric breakdown, which correctly reproduces the observed failure distribution.

- ESD device simulation can be applied to a wide variety of devices beside FETs, such as diodes, n well resistors, BJTs and SCRs. However, the calibration often has to be performed for each device type separately, owing to the space dependence of the SRH generation and recombination rates.

Bibliography

Abderhalden J., *Untersuchung zur Optimierung von Schutzstrukturen gegen Elektrostatis-che Entladungen in Integrierten CMOS-Schaltungen (in German)*, PhD thesis, Swiss Federal Institute for Technology, 1990.

Amerasekera A., Seitchik J., "Electrothermal Behavior of Deep Submicron nMOS Transistors under High Current Snapback (ESD/EOS) Conditions", Proc. IEDM (1994), 446.

Amerasekera A., Duvvury C., *ESD in Silicon Integrated Circuits, Second Edition,* John Wiley, Chichester, England, 2002.

Baliga B. J., *Modern Power Devices*, John Wiley, New York, 1987.

Bank R.E., Rose D.J. Fichtner W., "Numerical Methods for Semiconductor Device Simulation", IEEE Transact. Electr. Dev. **ED–30** (1983), 1031.

Bargstaedt S., Oettinger K., "Advanced 2D Latch-up Device Simulation — a Powerful Tool During Development in the Pre-silicon Phase", Proc. IRPS (2001), 235.

Benvenuti A., Ghione G., Pinot M.-R., Coughra W.-M, Schryer N.-L., "Coupled Thermal – Fully Hydrodynamic Simulation of InP Based HBTs", IEDM(1992), 737.

Bock K.-H., Keppens G., De Heyn V., Groesenecken G., Chiang L.Y., Naem A., "Influence of Gate Length on ESD Performance for Deep Submicron CMOS Technologies", Proc. 21st EOS/ESD Symposium (1999), 95.

Boselli G., Meeuwsen S., Mouthaan T., Kuper F., "Investigations on Double-diffused MOS Transistors under ESD Zap Conditions", Microelectronics Reliability, **41** (2001), 395.

Bufler F. M., Asahi Y., Yoshimura H., Zechner C., Schenk A., Fichtner W., "Monte Carlo Simulation and Measurement of Nanoscale n-MOSFETs", IEEE Trans. Elec. Dev. (2003).

Chen D., Yu Z., Wu K.-C., Grossens R.; Dutton R.W., "Dual Energy Transport Model with Coupled Lattice and Carrier Temperature", Proc. SISDEP (1993), 157.

Chynoweth A. "Ionization Rates for Electrons and Holes in Silicon", Phys. Rev. **109** (1958), 1537.

Chynoweth A. "Uniform Silicon p-n Junctions II. Ionization Rates of Electrons", J. Appl. Phys. **31** (1960), 1161.

Duvvury C., Amerasekera A., Joyner K., Ramaswamy S., Young S., "ESD Design for Deep Submicron SOI Technologies", Proc. VLSI Symposium Techn. (1996), 194.

DESSIS_ISE Reference Manual, ISE Integrated Systems Engineering AG, Zurich, Switzerland (1998).

ESD Standard Test Method for Electrostatic Discharge Sensitivity Testing — Human Body Model (HBM) Component Level (ESD STM5.1–2001), of *ESD Association Working Group WG 5.1*, 2001.

Wilkening W., "ESPRIT Project ESDEM", Proc. Public Workshop of the ESDEM, Institute for Integrated Systems, ETH Zurich (1999).

Esmark K., Fürböck C., Gossner H., Gross G., Litzenberger M., Pogany D., Zelsacher R., Stecher M., Gornik E., "Simulation and Experimental Study of Temperature Distribution During ESD Stress in Smart-Power-Technology ESD Protection Structures", Proc. IRPS (2000), 304.

Esmark K., *Device Simulation of ESD Protection Elements*, PhD thesis, Swiss Federal Institute for Technology, 2001; published Hartung-Gorre-Verlag, Konstanz, Germany, 2001.

Esmark K., Stadler W., Gossner H., Wendel M., Streibl M., Fichtner W., "ESD Schutzkonzeptentwicklung mit 2D- und 3D-Bausimulation - kann man auf Silizium verzichten (available in German only)", Proc. 7th ESD-Forum (2001a), 95.

Fong Y., Hu C., "The Effects of High Electric Field Transients on Thin Gate Oxide MOSFETs", Proc. 9th EOS/ESD Symposium (1987), 252.

Fong Y., Hu C., "High Current Snapback Characteristic for MOSFETs", IEEE Trans. Electr. Dev. **37** (1990), 2101.

Ghandi S. K., *Semiconductor Power Devices*, John Wiley, New York, 1978.

Groeseneken G., "Hot Carrier Degradation and ESD in Submicron CMOS Technologies: How do they Interact?", 22nd EOS/ESD Symposium (2000), 276.

Grutchfield H.B., Moutoux T.J., "Current mode second breakdown in epitaxial planar transistors", IEEE Trans. Electr. Dev. **11** (1966), 743.

Häcker R., Hangleiter A., "Intrinsic Upper Limits of Carrier Lifetime in Silicon", J. Appl. Phys. **75** (1994), 7570.

Hänsch W., *The Drift Diffusion Equation and Its Applications in MOSFET Modeling* , Springer, Wien, 1991.

Hirsch I., Berman E., Haik N., "Thermal Resistance Evaluation in 3-D thermal simulation of MOSFET transistors", Solid State Electronics **36** (1993), 106.

Ieong M., Tang T., "Influence of Hydrodynamic Models on the Prediction of Submicrometer Device Characteristics", IEEE Trans. Elec. Dev., **44** (1997), 2242.

Jacoboni C., Lugli P., *The Monte Carlo Method for Semiconductor Device Simulation*, Springer, Wien, 1989.

Lackner T., "Avalanche Multiplication in Semiconductors: A modification of Chynoweth's Law", Solid State Electr. **34** (1991), 33.

Maes W., de Meyer K., van Overstraten R., "Impact Ionization in Silicon: A Review and Update", Solid State Electr. **33** (1990), 705.

Miller J., Khazhinsky M., Weldon J., "Engineering the Cascoded NMOS Buffer for Maximum V_{t1}", Proc. EOS/ESD Symposium (2000), 308.

Mußhoff C., Wolf H., Egger P., Gieser H., Guggenmos X., "Risetime Effects of HBM and Square Pulses on the Failure Thresholds of GGNMOS Transistors", Proc. ESREF (1996), 1746.

Ning T. H., "Silicon Technology Directing in the new Millenium", Proc. IRPS (2000), 1.

Noetermans G., "On the Use of N-Well Resistors for Uniform Triggering of ESD Protection Elements", Proc. 19th EOS/ESD Symposium (1997), 221.

Okuto Y., Crowell C.R., "Threshold Energy Effects on Avalanche Breakdown Voltage in Semiconductor Junctions", Solid State Electr. **18** (1975), 161.

Overstraten van R., de Man H., "Measurements of the Ionization Rates in Diffused Silicon $p - n$ Junctions", Solid State Electr. **13** (1970), 583.

PARASITICS (Funding Project of the German Federal Ministry of Education and Science) "Simulation Statischer und Dynamischer Durchbruch (in German only)", Final Report (2000).

Perko L., *Differential Equations and Dynamical Systems*, Springer, New York, 1991.

Pierantoni A., Liuzzo A., Ciampolini P., Baccarani G., "Three-Dimensional Implementation of a Modified Transport Model", Proc. SISDEP (1993), 125.

Pogany D., Esmark K., Litzenberger M., Fürböck C., Goßner H., Gornik E., "Bulk and Surface Degradation Mode in 0.35 μm technology gg-nMOS ESD protection devices", Proc. ESREF (2000), 1467.

Russ, C., *ESD Protection Devices for CMOS Technologies: Processing Impact, Modeling, and Testing Issues,* PhD thesis, Technical University Munich, 1999; published Shaker-Verlag, Aachen, Germany, 1999.

Russ C., Mergens M.P.J., Verhaege K.G., Arner J., Jozwika P.C., Kolluri G., Avery L.R. "GGSCRs: GGNMOS Triggered Silicon Controlled Rectifiers for ESD Protection in Deep sub-micron CMOS Processes", Proc. 23rd EOS/ESD Symposium (2001), 22.

Salman A., Gauthier R., Furkay S., Muhammad M., Putnam C., Ioannou D., Nguyen P., Stadler W., Esmark K., "Characterization and Investigation of the Interaction Between Hot Electron and Electrostatic Discharge Stresses Using NMOS Devices in 0.13 μm Technology", Proc. IRPS, (2001), 219.

Schenk A., *Advanced Physical Models for Device Simulation*, Springer, Wien, 1998.

Shockley W., "Problems Related to p-n Junction in Silicon", Solid State Electr. **2** (1961), 35.

Seeger K., *Semiconductor Physics*, Springer, Berlin, 1991.

Selberherr S., *Analysis and Simulation of Semiconductor Devices*, Springer, Wien, 1984.

Slotboom J.W., van Dort M.J., Hurkx G.A.M., Klaase D.B.M., Kloosterman W.J., van Rijs F., Streutker G., Velghe D., "Physical Modeling and Simulation of Advanced Si-Devices – An Industrial Approach", Proc. ESSDERC (1993), 327.

Stadler W., Guggenmos X., Egger P., Gieser H., Mußhoff C., "Does the ESD Failure Current Obtained by Transmission Line Pulsing Always Correlate to Human Body Model Tests?", Proc. 19th EOS/ESD Symposium (1997), 366.

Stadler W., Esmark K., Reynders K., Zubeidat M., Graf M., Wilkening W., Willemen J., Qu N., Mettler S., Etherton M., Wolf H., Gieser H., Soppa W., De Heyn V., Natarajan M., Groeseneken G., Morena E., Stella R., Andreini A., Litzenberger M., Pogany D., Gornik E. Foss C., Nuernbergk D., Konrad A., Frank M., "Test Circuits for Fast and Reliable Assessment of CDM Robustness of I/O stages", accepted for publication 25th EOS/ESD symposium (2003).

Streibl M., Esmark K., Sieck A., Stadler W., Wendel M., Szatkowski J., Gossner H., "Harnessing the Base-Pushout Effect for ESD Protection in Bipolar and BiCMOS Technologies", Proc. 24th EOS/ESD Symposium (2002), 73.

Stricker A., *Technology Computer Aided Design of ESD Protection Devices,* PhD thesis, Swiss Federal Institute for Technology, 2000; published Hartung-Gorre-Verlag, Konstanz, Germany, 2000.

Sze S. M., *Physics of Semiconductor Devices, 2nd Edition,* John Wiley & Sons, New York, 1981.

Valdinoci M., Ventura D., Vecchi M. C., Ruda M., Baccarani G., Illien F., Stricker A., Zullino L., "Impact Ionization at Large Operating Temperatures", Proc. SISDEP (1999), 27.

Verhaege K., Roussel P. J., Groeseneken G., Maes H.E., Gieser H., Russ C., Egger P., Guggenmos X., Kuper F.G., "Analysis of HBM ESD Testers and Specifications Using 4th Order Lumped Element Model", Proc. 15th EOS/ESD Symposium (1993), 129.

Voldman S.H., Botula A., Hui D.T., Juliano P.A., "Silicon Germanium Heterojunction Bipolar Transistor ESD Power Clamps and the Johnson Limit", Proc. 23rd EOS/ESD Symposium (2001), 326.

Wachutka G.K., "Rigorous Thermodynamic Treatment of Heat Generation and Conduction in Semiconductir Device Modeling", IEEE Trans. CAD Integr. Circuits and Systems (1990), 1141.

Wachutka G.K., "Unified Framework for Thermal, Eelectrical, Magnetic and Optical Semiconductor device Modeling", COMPEL Vol 10 No 4 (1991), 311.

Wilkening W., Stadler W., Willemen J., Esmark K., Wolf H., Gieser H., "Investigations on Charged Device Model for On-Chip ESD Protection in the ASDESE Project", Proc. 7th ESD-Forum (2001), 87.

Wu J., Juliano P., Rosenbaum E.,, "Breakdown and Latent Damage of Ultra-Thin Gate Oxides under ESD Stress Conditions", Proc. 22th EOS/ESD Symposium (2000), 287.

Wunsch D.C., Bell R.R., "Determination of Threshold Failure Levels of Semiconductor Diodes and Transistors due to Pulse Voltage", IEEE Trans. Nucl. Sci. **15** (1968), 244.

Yu B., Wang H., Joshi A., Xiang Q., Ibok E., Lin M.-R., "15 nm Gate Length Planar CMOS Transistor", IEDM 2001, Washington DC, (2001).

Chapter 6

ESD circuit simulation

6.1 Field of application

Since the synthesis of the first integrated circuits, circuit simulation has been used to characterize the behaviour of the circuit under the specified operating conditions. Today, circuit simulation is an indispensable tool for the design of analogue circuits, such as I/O circuits.

Circuit simulation is based on a netlist, containing simplified models ("compact models") of the devices used. It must be emphasized that the models are *simplified*. In contrast to device simulation, the models for the devices, e.g. FET devices, bipolars, diodes, resistors, etc., are not intended to reflect exactly the microscopic physical behaviour of the devices, but to reproduce correctly the electrical characteristics in the operating range. Although the models of the devices are often based, at first glance, on standard physical equations and generic models, the physical meaning of the parameters in these equations, and even the use of the equations under the distinct constraints, are often questionable. In general, circuit simulation is one step away from physical reality in the device towards a higher level of abstraction. This must be kept into mind, particularly if compact models for ESD are analysed. However, without this abstraction a simulation of circuitries even with only several to several tens of devices is impossible.

ESD circuit simulation has gained significant importance during recent years. This has been the result mainly of the increasing complexity of today's state-of-the-art I/O circuits. On the one hand, as a consequence of the rapid decrease in the feature size of the processes, the density of the devices has increased dramatically, offering the circuit designer the realization of more complex circuits. On the other hand, new interface definitions (e.g., PCI-X, differential I/O stages as LVDS) require new circuit concepts and, very often, special ESD solutions under ESD critical constraints. A further aspect is the steadily decreasing operating voltage, making noise coupling between circuit blocks a crucial issue, and resulting in ICs with several de-coupled power supply domains requiring complex protection strategies. Even for experienced ESD engineers, the prediction of the impact of an ESD event on a circuit involving fifty or more devices is a very challenging and

time-consuming task. Device simulation is only applicable to simple circuits with
a couple of devices, and certainly not to the circuitries used in modern interface
solutions or even to a simulation of the whole pad ring and some "parasitic" core
devices. Thus, designers will benefit from a convenient compact simulation tool
that can verify their ESD protection circuits through modelling the interaction
between I/O buffers and the pad circuitry.

In analogy to device simulation, the standard models used in conventional circuit simulators are initially not suitable for application in the ESD relevant regime
with high current densities and fast transients. Standard models for MOS transistors are in general unable to reproduce the avalanche breakdown correctly, not to
mention the snapback or the high-current behaviour. Diode models implemented
in circuit simulators are designed to match a diode's characteristic typically up to
a maximum of 1 mA/µm^2, and are, therefore, inappropriate for the simulation of
ESD events with current densities in the 10–100 mA/µm^2 regime. For this reason, dozens of new models or model extensions have been discussed in literature in
recent years.[1] Compact models suitable for ESD conditions exist for all possible
kinds of devices in all kinds of processes (see Section 6.2).

Clearly, the complexity and physical content of the compact models is inversely
proportional to the complexity of the circuitries which can be simulated with these
compact models. The limitation is the numerical stability and the computing speed
of today's computer systems and algorithms. For a "simplified" I/O circuit consisting of a buffer and a few ESD protection elements, advanced compact models
can be used to reproduce the static and dynamic behaviour of the device under
ESD stress rather accurately, including thermal coupling. In contrast, simulation
of a pad ring with hundreds of I/O cells will require a very slim model, concentrat-
ing only on the most important features. Such a slim model will describe the real
behaviour of the device only very roughly, regardless of whether it is used under
normal operating conditions or under ESD stress. The ESD engineer must always
strike a balance between physical and numerical requirements.

Recently, the complexity of the problems able to be solved with ESD circuit
simulation has steadily increased (see e.g. Puvvada 1998; Wolf 2001). Besides
HBM and MM stress, the response of circuits to CDM stress has been analysed
with circuit simulation (see e.g. Beebe 1998; Mergens 2000).

The aim of this chapter is not to provide a complete description of all the
compact models developed in recent years, but rather to give the essential concepts
of compact models applicable to ESD circuit simulation. With this introduction
and the many excellent references found in recent literature, the reader should
be able to set-up her/his own compact models relevant to the device and the
problem under consideration. As an example, the most commonly used models for
a FET, and the physical/electrical equations they are based on, are discussed for
a modern sub-µm process in Section 6.4. The parameter extraction and validation
of the compact model plays an important role for the successful use of the ESD
compact models (Section 6.4.3). In Section 6.5 several typical applications of ESD
compact models are discussed. These examples demonstrate the power of the

---

[1]The proceedings of recent EOS/ESD Symposia, IRPS, and ESREF are a rich source of
information on those models.

ESD circuit simulation, but also its limitations arising from the higher level of abstraction (Section 6.6).

6.2 Compact models for analogue simulation

In recent years, a large variety of compact models has been proposed for almost all possible protection devices and active elements. Even in a book dedicated to ESD simulation methods, it is impossible to discuss all conceivable models in detail. Sophisticated models are available for relatively "simple" elements, such as diodes where the model basically consists of two diffusions (e.g. Wang, 2000), but also for more complex structures where several possible current paths are competing (see, for example, the investigations of lateral DMOS power devices by Mergens 1999). Depending on the task and the devices used in the circuitry, the ESD engineer has to choose the most appropriate models. Extensions may be required to existing models, to account for effects which are observed in a particular process or device and influence the ESD behaviour. In such cases, detailed investigations, perhaps by means of DC and square pulsing techniques, are inevitable if one is to understand the device physics and implement the effects into the compact models in a physically meaningful way. Textbooks dealing with device physics, such as Sze (1981, 1997) and Muller (1986), are valuable sources for finding physically based model equations for the problem under consideration.

Diodes, resistors, and bipolar transistors (BJTs) are often included in ESD protection concepts, either as active ESD devices or as parasitic elements, inherently included in protection elements, drivers, guard rings, etc. These devices can be regarded as basic elements of the ESD circuit simulation. More "complex" components, such as ESD models of FETs, can often be assembled from these three devices with only minor extensions. The basic compact models and physical equations and concepts behind these models are discussed in the following sections.

In CMOS technologies, FETs are the workhorse for any kind of circuitry. In almost all conceivable I/O interface circuits in CMOS technologies, FETs are used as active elements, either in driver stages or in input inverter stages. Used as I/O drivers, the FETs are connected directly to the pad or directed to the pad via a small serial resistor, and, thus, a distinct portion of the discharge current can potentially flow across these devices. In addition, FETs are often used in ESD protection concepts, either as grounded-gate FETs with the gate tied to ground or with the gate coupled to VDD and VSS. Therefore, in CMOS technologies, ESD circuit simulation would be unthinkable without appropriate models for FETs covering the ESD-relevant effects. It is therefore not surprising that most of the work found in the literature is dedicated to the behaviour of FETs under ESD stress. Because of the outstanding relevance of FETs to ESD protection concepts, the modelling of FETs forms the main part of this chapter.

Besides the three basic elements, diodes, resistors and BJTs, and FETs, some compact models exist for special protection devices. In the literature, compact models of SCRs (Li, 1997; Juliano, 2001) and of a DMOS in smart-power technology (Mergens, 1999) are discussed. These devices will not be treated here; the

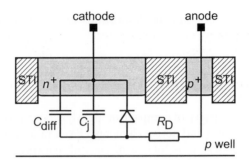

Figure 6.1: Schematic cross section of a diode in an STI technology with one possible compact model. The compact model includes the junction and diffusion capacitance, the serial resistance, and the diode characteristic itself. The serial resistance may depend on the current density. Possible physical mechanisms included in the diode model are described in the text.

reader is referred to the original literature in which these compact models are explained and validated in detail.

6.3 Basic compact models of protection devices

6.3.1 Diodes

In all probability, diodes will be included in all possible ESD protection concepts. Because of their excellent shunt behaviour, they are often used under forward bias conditions as protection elements. However, in standard CMOS processes an n driver implies a parasitic n^+p diode from the n^+ diffusion connected to the pad to the p well connected in general to VSS. In analogy, a p driver in a non-overvoltage tolerant application contains a parasitic p^+n diode from pad to VDD. The cross-sectional view of a diode together with one possible compact model is depicted in Figure 6.1.

Diodes are the simplest bipolar structure in a process and therefore have for a long time been analysed theoretically and experimentally. The ideal IV characteristic of diodes is given by Shockley's law (Shockley, 1950)

$$J(V) = J_s \left\{ \exp\left(\frac{qV}{k_B T}\right) - 1 \right\} \quad \text{with} \quad J_s = \frac{q n_{p0} D_n}{L_n} + \frac{q p_{n0} D_p}{L_p} \qquad (6.1)$$

where k_B is Boltzmann's constant, and T is the temperature which determines the exponential characteristic, by which at room temperature a voltage change of ≈ 60 mV causes an increase of the current density of one decade. The reverse saturation current density J_s depends on material-related parameters, namely the diffusion coefficients for electrons and holes D_n and D_p, the diffusion lengths L_n and L_p and the equilibrium electron density on the p side n_{p0} and the equilibrium hole density on the n side of the diode p_{n0}.

Strictly speaking, Shockley's law is only valid under very restricted conditions (Sze, 1981). Among other things, the theory presumes abrupt junctions, an abrupt

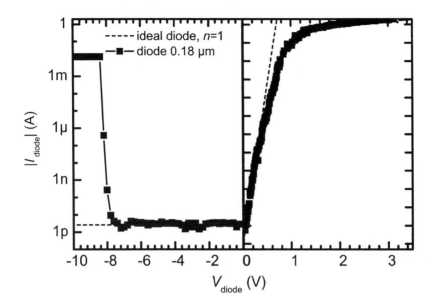

Figure 6.2: DC measurement of a p^+n diode in a 0.18 µm CMOS process compared to the ideal Schottky diode characteristic according to Equation 6.1. In reverse mode, the compliance setting limits the maximum current, in order to avoid thermal destruction of the diode.

depletion layer and electrical neutrality outside the depletion region and low injection, and it neglects the generation/recombination currents in the depletion zone and any series resistances. Therefore it is not surprising that this simple theory cannot describe diodes in modern processes for a large range of bias conditions. A comparison of experimentally obtained data in a modern 0.18 µm CMOS process and Shockley's law yield an acceptable consistency only in the range of low current densities (see Figure 6.2). Neither the breakdown behaviour under reverse-biased conditions nor the forward-biased high-injection regime can be modelled adequately.

For high-injection under forward-biased conditions, the drift current arising from the electric field in the neutral regions caused by the excess carriers must be considered. To a first approximation (Sze, 1981), the exponent of Equation 6.1 changes from qV/k_BT to $qV/2k_BT$ (or, more generally, to qV/nk_BT with the ideality factor/emissivity $n = 1\ldots2$ depending on the level of injection. The transition from low injection ($n = 1$) to high injection ($n = 2$) can be accounted for by using

$$J(V) = \frac{J_{\text{ideal}}}{1 + \sqrt{\frac{J_{\text{ideal}}}{J_{\text{hi}}}}} \tag{6.2}$$

where J_{ideal} is the ideal diode IV characteristic, and the fitting parameter J_{hi} describes the transition from $n = 1$ to $n = 2$. Models according to Equation 6.2 can

easily be implemented in circuit simulators (Massobrio, 1993). Alternatively, an
analytical solution for the IV characteristic under high-injection level conditions
can be calculated (Wang, 2000) using the junction built-in potential and the ratio
of the doping concentration $N_{\mathrm{D}}/N_{\mathrm{A}}$ and $N_{\mathrm{A}}/N_{\mathrm{D}}$ as fitting parameters.

Clearly, for *real* diodes, a serial resistance which might be conductivity modulated must be taken into account. A detailed analysis is given by Russ (1999).
The serial resistance R_{D} causes an additional voltage drop. The voltage V across
the *real* diode (see Figure 6.3) reads as

$$V(J) = \ln\left(\frac{J_{\mathrm{ideal}}}{J_{\mathrm{s}}}+1\right)\frac{nk_{\mathrm{B}}T}{q} + AJR_{\mathrm{D}} \tag{6.3}$$

$$\text{with}\quad J_{\mathrm{ideal}} = \left(\frac{J}{2\sqrt{J_{\mathrm{hi}}}}-\sqrt{\frac{J^2}{4J_{\mathrm{hi}}}+J}\right)^2$$

where A is the *pn* junction area. For very high current densities, such as during an
ESD event, R_{D} becomes a function of the current density itself. For the dependence
of R_{D} on the current density through the resistor or the voltage across the serial
resistance, several approximations exist with more or less physical background and
with different numerical approaches (Russ, 1999; Puvvada, 2000; Wang, 2000).

Equation 6.1 does not include any breakdown mechanism under reverse-biased
conditions. However, in an ESD-suitable compact model, the breakdown behaviour
of the diode must be included. Wang (2000) obtained excellent agreement between
simulation and experiment by using the expression

$$J_{\mathrm{rev}} = J_{\mathrm{rev0}}M\left(1-\mathrm{e}^{-C_{\mathrm{i}}V}\right) \tag{6.4}$$

with the Miller multiplication factor M (see Sze 1981 and Section 6.3.3) and the
fitting parameters J_{rev0} and C_{i}.

Often, the detailed breakdown behaviour is not required because of the very
bad current conducting capability of the diode under reverse-biased conditions. In
this case, a passable solution could be to use a simplified model incorporating a
Zener diode model anti-parallel to the diode. A warning flag must be set by the
circuit simulator in case the Zener diode turns on. A more sophisticated model
is discussed by Russ (1999) and Mergens (2001) which takes into account the
mobility degradation of the carrier under high reverse injection.

Wolf (1996) and Wang (2000) have included thermal self-heating effects in their
diode compact models. Although for standard bulk diodes in a CMOS process
sufficient agreement with the experimental data can be obtained without including
self-heating effects, for special applications with reduced heat flow, such as diodes
in SOI technologies (Wang, 2000), self-heating must be considered. The problems
of the advanced diode models are reduced numerical stability and a complex model
dependence on process and layout parameters.

Finally, to account for the dynamic behaviour of the diode, the junction ca-
pacitance C_{j} and the diffusion capacitance C_{diff} must be included in the model.
To a first approximation, the voltage dependence of the junction capacitance (Sze,
1981) can be neglected.

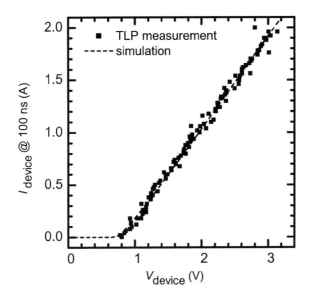

Figure 6.3: Measurement and compact simulation of the high-current characteristic of a p^+n diode in a 0.18 μm CMOS process. The experimental data are obtained by 100 ns transmission line pulsing.

A typical example for the forward bias IV characteristic of a p^+n diode in a 0.18 μm CMOS technology is given in Figure 6.3. For the model, a network according to Figure 6.1 has been used, and the diode was modelled with Equation 6.3. The rather simple source code of the diode, written in VHDL–AMS, is listed in Figure 6.4. The simulation using the compact model matches experimentally obtained values pretty well. For this particular diode, it turns out that the dominant parameter determining the IV characteristic is the serial resistance R_D.

It should be noted here that the correlation between experimental results and simulation using the diode model discussed above is much worse if very fast pulses, i.e. pulses in the CDM time domain, are applied to the diode. A possible explanation is that the very short pulses cannot generate enough carriers to flood the diode and, therefore, the diode is comparably high resistive. Consequently, for analysis in the CDM time domain, further extensions of the diode compact models have to be developed. An example for a possible extension is the Ma-Lauritzen model which includes the so-called forward-recovery effect (Ma, 1997).

6.3.2 Resistors

Resistors are commonly used in ESD protection concepts (see Chapter 2). Besides their use as a de-coupling element between the primary ESD protection stage and the driver or input gate node, resistors are often required by the I/O circuit design itself, e.g. for matching purposes or for the termination of rf circuits. In both cases, the resistors are potentially in the discharge path and, therefore, must be

```
-----------------------------------------------------------------------
-- File: diode.vh;  2002-10-18 (c) Stephan Drueen
-----------------------------------------------------------------------


PACKAGE electrical_pack IS
        -- declare subtypes for voltage and current, with units
        subtype voltage is real;
        subtype current is real;
        subtype charge is real;
        -- basic nature and reference terminal for ELECTRICAL systems
        nature ELECTRICAL is
                voltage across
                current through
                ground  reference;
END electrical_pack;


-----------------------------------------------------------------------
ENTITY diode  IS
        GENERIC (  Ad      : real := 1.0e-12   ;
                   Ihi     : real := 1.0e-3    ;
                   Iss     : real := 1.0e-6    ;
                   N       : real := 1.0);
        PORT(TERMINAL a, c : electrical);
END ENTITY diode  ;
-----------------------------------------------------------------------


ARCHITECTURE nppdiode OF diode IS

        CONSTANT kb : real := 1.38066e-23;
        CONSTANT q  : real := 1.60218e-19;
        CONSTANT T  : real := 300.0;
        CONSTANT vt : real := 26.0e-3;
        CONSTANT c1 : real := N*kb*T/q;

        -- Controlling voltages
        QUANTITY Vd ACROSS a TO c;
        QUANTITY Id THROUGH a TO c;

        -- Variables
        QUANTITY iideal : real;

        BEGIN
        iideal == Iss*(exp(Vd/c1)-1.0);
        Id == iideal/(1.0+sqrt(realmax(iideal/Ihi,1.0e-8)));
END nppdiode ;
```

Figure 6.4: VHDL–AMS source code of the diode used in Figure 6.1. The diode *IV*
characteristic is modelled with Equation 6.2.

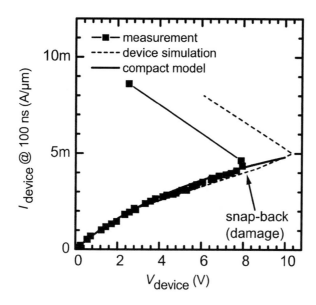

Figure 6.5: Measurement and compact simulation of the high-current characteristic of an n-well resistor in a 0.13 μm CMOS process. The experimental data are obtained by 100 ns transmission line pulsing. For comparison, the device simulation is included. Except for the breakdown behaviour, experiment and simulations show very good agreement.

designed and modelled accordingly.

A typical IV characteristic of an n well resistor is shown in Figure 2.22. An experimentally obtained characteristic is depicted in Figure 6.5. The ideal linear ohmic behaviour can only be obtained in a range of rather low current density. For most circuit designs, this range is the one used under standard operating conditions. Hence, standard models for resistors in circuit simulators often cover only this linear range.

For ESD designers, current densities far above the defined operating conditions are of interest. For high current densities, a remarkable increase in the resistance can be observed for diffusion resistors (see Section 2.4.5 and Section 5.5.2). Of course, this resistance increase can be exploited by the ESD designer (e.g., Carbajal, 1992). At even higher current densities, snapback occurs, and the resistor may be irreversibly damaged. The exact IV characteristic depends on the current density and the layout of the resistor.

The current through a one-dimensional homogeneous n-type resistor is in general given by (Sze, 1981)

$$\vec{J}_{\mathrm{n}} = q\vec{v_{\mathrm{d}}}n \tag{6.5}$$

where $\vec{v_d}$ is the electron drift velocity and n is the electron density. For p-type resistors the equations are in analogy. For simplicity, the resistor is regarded as a one-dimensional problem, so Equation 6.5 reads as $J_{\mathrm{n}} = q v_{\mathrm{d}} n$. Any contributions to the current by diffusion are neglected. At low electric fields, and therefore low voltages across the effective resistances, the drift velocity depends linearly on the electric field $\vec{E}$

$$\vec{v_\mathrm{d}} = \mu_\mathrm{n}\vec{E},\qquad(6.6)$$

or $v_\mathrm{d} = \mu_\mathrm{n}E$ for 1D problems where the proportionality factor μ_n (assumed to be isotropic) is the mobility of the electrons. In silicon, mobility is limited by the scattering of electrons with acoustic phonons and impurities. The latter is, to a first approximation, inversely proportional to the donor concentration N_D (Kittel, 1983).

With increasing carrier energies in high electric fields, the scattering on optical phonons must be considered and the effective mobility is decreased (Sze, 1981). The drift velocity for a one-dimensional problem can be estimated as

$$v_\mathrm{d} = \mu_0 E\sqrt{\frac{T}{T_\mathrm{e}}} \text{ with } \frac{T}{T_\mathrm{e}} = \left\{1 + \left(1 + \frac{3\pi\mu_0^2 E^2}{8c_s^2}\right)^{1/2}\right\}\qquad(6.7)$$

where T_e is the effective temperature of the carriers, T is the lattice temperature, μ_0 is the low-field mobility and c_s is the velocity of sound. Finally, for $\mu_0 E \gg c_\mathrm{s}$, the drift velocity saturates to

$$v_\mathrm{sat} = \sqrt{\frac{8E_\mathrm{op}}{3\pi m_0}} \approx 10^5\,\frac{\mathrm{m}}{\mathrm{s}}\qquad(6.8)$$

where E_op is the energy of optical phonons and m_0 is the effective electron mass in silicon.

Equation 6.7 can be solved analytically. However, to a first approximation, an "effective mobility" can be defined over the whole range of the drift velocity, leading to an expression for the current density in the resistor of

$$J_\mathrm{n} = qn\frac{\mu_0 v_\mathrm{sat}}{v_\mathrm{sat} + \mu_0 E}E.\qquad(6.9)$$

With homogenous doping in the resistor, the electric field over the resistor can be estimated to be constant, i.e., $E = V/L$ with the resistor length L. Both Equations 6.7 and 6.9 describe the experimentally obtained data quite well (see Figure 6.5, where Equation 6.9 is used to fit the IV characteristic of an n-well resistor).

In principle, the parameter extraction procedure is straightforward; in the "usual" range of dimensions for an ESD-relevant application of the resistor, scaling can be obtained. However, the parameter extraction for the model using Equation 6.9 is not as easy as it seems at first glance. Saturation velocity and mobility depending on the doping concentration can be found in textbooks. However, choosing the electron concentration n is more tricky. Although the doping concentration of the resistor is in general known from the process information, a complete doping profile is required in order to find an appropriate mean value of the doping concentration which fits the cross section, i.e. the depth of the resistor and the current flow. A possible solution is to fractionalize the doping profile along its depth, until the whole profile is divided into slices with approximately constant doping concentrations. The total resistor can then be taken as a parallel circuit of resistances represented by the slices. However, in most cases the effective doping concentration and the corresponding area of the cross section of the resistor may simply be regarded as fitting parameters to match circuit simulation with device simulation and/or experiment.

When the electric field in the resistor is high enough that the carriers can gain sufficient energy to generate further electron–hole pairs, avalanche generation occurs (see Section 5.5.2). Avalanche generation can be accounted for by including a multiplication factor M in Equation 6.9 (Puvvada, 2000). However, for the resistors modelled in Figure 6.5, a snapback results in immediate and irreversible damage to the device, and therefore for this special application snapback effects itself are not considered. Finally, the question is whether the heat dissipation in the resistor must be included in the model. Basically, all parameters in the Equations 6.5–6.9 are temperature dependent. Furthermore, self-heating effects are known to exist in highly-doped resistors (Amerasekera, 1993). However, given the good agreement between experiment and simulation with the models disregarding heating effects (see Figure 6.5), further extensions to these models seem not to be necessary. For special applications, e.g. resistors with very small width and under high stress levels, the temperature increase under ESD stress must be included in the model.

6.3.3 Bipolar junction transistors

The bipolar junction transistor (BJT) is undoubtedly the most important single device for ESD modelling purposes. Figure 6.6 gives some examples for BJTs formed in a process. First, a concrete vertical or lateral BJT can be processed to be used as a stand-alone ESD protection element. Second, in bipolar technologies, BJTs are the workhorse of the designer. Furthermore, parasitic BJTs are omnipresent and unavoidable in each circuit, e.g. as the substrate pnp of a PFET in a CMOS technology. Therefore a detailed understanding of the BJT action under ESD stress is essential for successful circuit simulation. In this section we will discuss the modelling of an npn BJT. Other bipolar transistors can be modelled in analogy. The basic ideas for modelling BJTs hold for all types of the above mentioned type of BJTs in a circuit, although obviously the parameter extraction methodology is quite different.

The typical IV characteristic of a BJT can be divided into three parts (see Figure 6.7). The first part of the characteristic describes the behaviour under standard operating conditions, within the specified range for the applied voltages and the currents through the BJT. Usually, this regime is well modelled by standard circuit simulation models, such as used in SPICE. The second part of the IV characteristic is called the turn-on and snap-back regime; the third part describes the high-current characteristic of the BJT after turn-on. Part II and part III of the IV characteristic are governed by high-current effects, which are dominant under ESD stress. For this regime extensions of the compact model beyond the standard models are required.

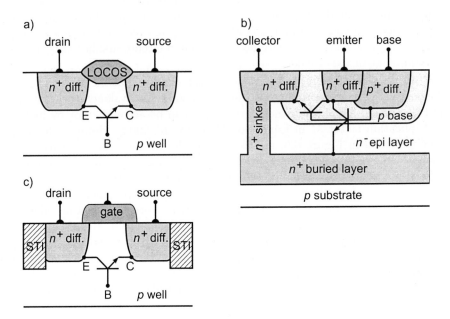

Figure 6.6: Examples of realizations of BJTs in different technologies. (a) Thick field-
oxide device. (b) Lateral and vertical *npn* transistors in a bipolar technology.
(c) Parasitic *npn* BJT of an NFET in an STI technology.

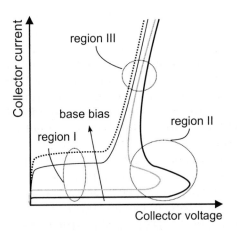

Figure 6.7: Schematic representation of the *IV* characteristic of a BJT with different base
bias. Region I: linear and saturation characteristic; region II: breakdown and
snapback characteristic; region III: high-current characteristic.

Operation in the standard regime

At the circuit level, BJT operation in the *standard* operating regime can be modelled quite accurately for many applications by the well-established Gummel–Poon model (Gummel, 1970). In the Gummel–Poon (GP) model, the terminal currents and the junction voltages are linked to the base charge Q_B which is a sum of five components

$$Q_B = Q_{B0} + Q_{jE} + Q_{jC} + Q_{dE} + Q_{dC}. \tag{6.10}$$

where Q_{B0} is the base charge under zero bias, and $Q_{jC} = C_{jC}V_{CB}$ and $Q_{jE} = C_{jE}V_{EB}$ are voltage dependent charges related to the depletion region of emitter and collector, respectively. The charges Q_{dE} and Q_{dC} are the charges stored in the diffusion capacitances and are related to the current in forward direction I_F and reverse direction I_R by

$$Q_{dE} = B\tau_F I_F \text{ and } Q_{dC} = \tau_R I_R \tag{6.11}$$

where τ_F is the lifetime of the minority carriers in the forward direction and τ_R is the equivalent in the reverse direction. The factor B accounts for the Kirk effect. The base current is given by

$$I_B = \frac{dQ_B}{dt} + I_{\text{rec}}. \tag{6.12}$$

In the GP model, the base recombination current is usually modelled by two contributions $I_{\text{rec}} = I_{EB} + I_{CB}$ with diode-type characteristics

$$I_{EB} = I_{S1}\left\{\exp\frac{qV_{EB}}{k_BT} - 1\right\} + I_{S2}\left\{\exp\frac{qV_{EB}}{n_Ek_BT} - 1\right\} \tag{6.13}$$

$$I_{CB} = I_{S3}\left\{\exp\frac{qV_{EB}}{n_Ck_BT} - 1\right\} \tag{6.14}$$

The above considerations can be summarised in an equivalent circuit model for the BJT (Figure 6.8). This circuit gives a highly accurate description of the BJT, including many physical effects such as current-induced base push-out, or Kirk effect (factor B in diffusion capacitance Q_{dE}), and the Early effect (a voltage dependence of Q_{jC}). Consequently, the GP model is a standard model in current circuit simulators. The down side is that the accuracy and the physical models can only be achieved by using a large number (> 25) of parameters, all of which must be extracted from measurements. Very often the modelling of the detailed physical effects in the standard operating regime is not the ESD engineer's focus, and the parameter extraction procedure for these effects can be rather time consuming and complicated. Furthermore, the use of the GP model in the ESD simulation often leads to numerical instabilities, owing to the large number of equations and elements. Therefore, a simplified model with reduced physical contents is often used in ESD circuit applications.

For ESD purposes, the classical Ebers–Moll (EM) model is, in the authors opinion, an acceptable compromise between complexity and physical contents on the one side and handling and numerical stability on the other side. The EM

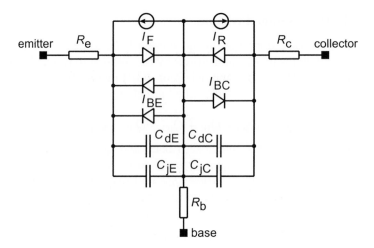

Figure 6.8: Equivalent circuit for the Gummel–Poon model according to Gummel (1970).
The two parallel diodes account for the two components of I_{BE} with ideality
factors 1 and n_{E}.

model is widely used as the basis for ESD circuit simulation of either parasitic or
concrete bipolar devices, see, for example, Luchies (1994), Russ (1996), Stricker
(1998), Wolf (2001).

In its original form, the EM model is designed to describe the large-signal
behaviour of BJTs (Ebers, 1954). The basic equivalent circuit consists of two
back-to-back connected diodes, accounting for the collector–base and the base–
emitter diode, and two current sources which depend on the currents through the
parallel diodes (see Figure 6.9 a)). For simulation purposes, the transport version
of the EM model is mostly used (Massobrio, 1993). In the basic equivalent circuit
of the transport version, the currents through the current sources are given by

$$I_{\mathrm{cc}} \;=\; I_{\mathrm{sat,be}} \left\{ \exp\left(\frac{qV_{\mathrm{be}}}{n_{\mathrm{be}}k_{\mathrm{B}}T} \right) - 1 \right\} \tag{6.15}$$

$$I_{\mathrm{ec}} \;=\; I_{\mathrm{sat,bc}} \left\{ \exp\left(\frac{qV_{\mathrm{bc}}}{n_{\mathrm{bc}}k_{\mathrm{B}}T} \right) - 1 \right\}. \tag{6.16}$$

where $I_{\mathrm{sat,be}}$ and $I_{\mathrm{sat,bc}}$ are the saturation currents for the base–emitter diode
and the base–collector diode, respectively, and n_{be} and n_{bc} are the corresponding
ideality factors. The terminal currents flowing into the model are then given by
(see Figure 6.9 a))

$$I_{\mathrm{c}} \;=\; I_{\mathrm{cc}} - \frac{I_{\mathrm{ec}}}{\alpha_{\mathrm{r}}} \tag{6.17}$$

$$I_{\mathrm{e}} \;=\; I_{\mathrm{ec}} - \frac{I_{\mathrm{cc}}}{\alpha_{\mathrm{f}}} \tag{6.18}$$

$$I_{\mathrm{b}} \;=\; \left(\frac{1}{\alpha_{\mathrm{f}}} - 1 \right) I_{\mathrm{cc}} + \left(\frac{1}{\alpha_{\mathrm{r}}} - 1 \right) I_{\mathrm{ec}} \tag{6.19}$$

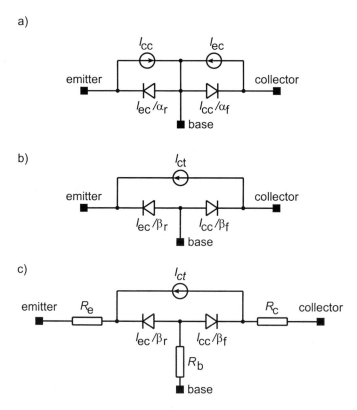

Figure 6.9: a) Equivalent circuit for the Ebers–Moll model according to Ebers (1954) (transport version). b) Transport version with a single current source. c) Extended model including terminal resistors.

α_f and α_r denote the forward and reverse current gain of a common-base BJT which can be expressed by the forward and reverse current gains of a common-emitter transistor by

$$\alpha_f = \frac{\beta_f}{1 + \beta_f} \quad \text{and} \quad \alpha_r = \frac{\beta_r}{1 + \beta_r} \tag{6.20}$$

In symmetric BJTs, the simplification $\beta_r = \beta_f = \beta$ is sound. β is proportional to the emitter doping N_e and inversely proportional to the Gummel number Q_b, representing the total number of impurities in the base (Sze, 1981).

If the diode characteristics of the emitter–base and the collector–base junction are assumed to be ideal, i.e. $n_{be} = n_{bc} = 1$, in the set of equations 6.15–6.19 only four parameters exist to be determined from measurement or calculation. These are the saturation currents $I_{sat,be}$ and $I_{sat,bc}$ and the current gains α_r and α_f.

Commonly, the two current sources I_{cc} and I_{ec} are merged to one single current source I_{ct} (Figure 6.9 b)) with

$$I_{ct} = I_{cc} - I_{ec} \tag{6.21}$$

To improve the accuracy of the basic EM model, terminal resistors R_b, R_c, and R_e are included (see Figure 6.9 c)). Often, the static behaviour of the BJT can be sufficiently well explained by the EM model of Figure 6.9 c).

Avalanche multiplication

With high-injection levels, additional effects occur which are not included in the aforementioned GP or EM models. Consider a positive ESD stress applied to the collector of the BJT in Figure 6.6, while the emitter is grounded. Initially, the BJT is in a state of high impedance. Because of the high potential at the collector and the resulting high electric field at the collector–base junction, holes are injected into the base region. If the electric field is sufficiently high, the energy of the holes will be high enough to generate further electron–hole pairs by impact ionization, and an avalanche effect is triggered. The holes drift towards the base, causing a voltage drop across the parasitic base resistance. If the local potential at the emitter–base junction exceeds approximately 0.7 V, the *npn* turns on, the voltage across the BJT breaks down, and the *IV* curve shows a negative resistance with a characteristic snapback. Increasing the current further leads to an increased voltage drop arising from conductivity modulation in the base resistance and the parasitic series resistances in the emitter and collector area.

It is obvious that neither the GP model nor the EM model can reproduce the behaviour of a BJT under ESD stress. Neither avalanche generation and breakdown, nor turn-on of the bipolar under ESD currents, is considered in these models. A common approach for ESD simulation is to extend the models by current sources. The purpose of these current sources is to describe the physical effect of avalanche multiplication as exactly as possible. An appropriate implementation of the avalanche multiplication current sources with reasonable physical models is crucial for the predictability and numerical stability of the model.

The most important parameters and models are discussed in terms of the model of Figure 6.10. In addition to the standard EM model discussed above, a current source I_{avc} has been included to account for the avalanche generation between collector and base. Furthermore, an internal base resistance R_w has been included which, in general, is current dependent. The model can be further generalized by adding an avalanche current source I_{ave} to account for possible avalanche multiplication between emitter and base. The avalanche current I_{avc} resulting from the impact ionization of the holes injected into the base can be written simply as (Dutton, 1975)

$$I_{avc} = (M - 1)I_{ct} \tag{6.22}$$

where M denotes the multiplication factor; in fact, here it is the multiplication factor for holes. The multiplication factor is defined as the ratio of the current at the beginning and at the end of the depletion region. Assuming equal ionization coefficients for electrons and holes $\alpha_n \approx \alpha_p \approx \alpha$ (which is roughly justified in silicon), the correlation between M and the ionization coefficient is (Sze, 1981)

$$1 - \frac{1}{M} = \int_0^W \alpha dx \tag{6.23}$$

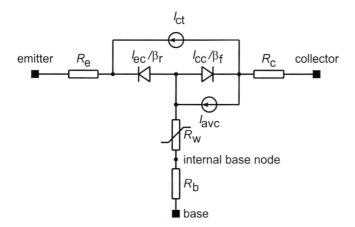

Figure 6.10: EM model extended by an avalanche multiplication current source and a current dependent base resistance. Similar models are commonly used for ESD circuit simulations of BJTs.

where W is the width of the depletion region. Elaborate theories exist for α. For the most part, Chynoweth's approximation for the dependence of the ionization coefficient on the electric field across the junction (Chynoweth, 1960)

$$\alpha = a \exp\left(\frac{-b}{E}\right) \tag{6.24}$$

leads to reasonable results. In the above equation, a and b are fitting parameters. For high fields, such as in the case of snapback, the electric field across the junction can be assumed to be constant and may be expressed by the ratio of the voltage across the junction, V_j and the width of the junction W, $E = V_j/W$. Equation 6.23 and 6.24 combine to give:

$$M = \frac{1}{1 - \alpha W} = \frac{1}{1 - aW\exp\left(-bW/V_j\right)} \tag{6.25}$$

Alternatively, the empirical Miller formula (Miller, 1957) is often used. This links the multiplication factor with the voltage across the junction V_j and the breakdown voltage BV_{CB0} of the collector–base diode (open emitter):

$$M = \frac{1}{1 - \left(\frac{V_j}{BV_{CB0}}\right)^n} \tag{6.26}$$

The exponent n can be used as a fitting factor in the range of 2–6. Wolf (1998) showed that for sub-μ technologies in which part of the high-current IV characteristic is at rather low voltages, M can be better modelled by using a voltage dependent n.

There are two possibilities for extracting the model parameters in the avalanche regime. First, the dependence of the multiplication factor on the voltage across the junction can be explicitly measured with a three-terminal device using $M = I_c/(I_c - I_b)$ as derived from Equation 6.22, under the assumption that before the BJT is turned on, the entire generated current is flowing into the base. Second, the avalanche breakdown voltage can be measured. At the breakdown of the junction, $M \rightarrow \infty$, and consequently, $V_j = bW/\ln(aW)$ from Equation 6.25.

Turn-on of a BJT under ESD stress

For analysis of a BJT under ESD stress, the turn-on of the bipolar transistor is of major importance. The turn-on of the bipolar, either parasitic or as a concrete element, is utilized in many different ESD concepts. The triggering of a BJT causes a negative resistive branch in the IV characteristic and a voltage snapback. Because of the voltage snapback, the power dissipation in the device is reduced, increasing the ESD robustness, assuming a homogeneous turn-on. One condition for turn-on of the BJT defines the local voltage which is required to forward bias the emitter–base junction

$$V_{be} = I_b R_b \geq \sim 0.7 \text{ V} = V_{on} \tag{6.27}$$

The value of 0.7 V is a rough estimation; in general, values between 0.5–0.9 V are appropriate. A second condition for turn-on is derived from the current analysis of the EM model including an avalanche source according to Figure 6.10 (see e.g. Amerasekera (1996)). As the BJT is turned on, Equation 6.22 can be written as

$$I_b + \frac{I_{cc}}{\beta} = (M - 1)I_c \tag{6.28}$$

With $I_c/I_b = \beta$ and the assumption that $I_{cc} > 0$, Equation 6.28 leads to the well-known expression

$$\beta(M - 1) = k \geq 1 \tag{6.29}$$

The coefficient k can be used to tune the snapback behaviour. Remember that the multiplication factor M is voltage dependent according to Equation 6.23 or 6.25. As the base–emitter junction is forward-biased and the bipolar is turned on, the emitter current feeds additional carriers into the avalanche source. Consequently, the voltage required to keep the bipolar in the on-state, so as to fulfil Equation 6.29, can be decreased. The voltage snaps back to the holding voltage V_h, which is sometimes called the sustaining voltage or (somewhat misleadingly) the snapback voltage. V_h is the smallest voltage at which bipolar action can be sustained. Its value can be estimated by using Equation 6.26, 6.27 and 6.29.

$$V_h \approx BV_{CB} \sqrt[n]{\frac{k}{k + \beta}} + V_{on} \tag{6.30}$$

The holding voltage depends on the gain of the BJT, β, the multiplication factor, M, and on the collector–base breakdown voltage BV_{CB}. In the turn-on state of the bipolar, the terminal resistors R_e and R_c must be included. These resistors lead to an additional voltage drop $V = I_{device}(R_c + R_e)$, especially at high current

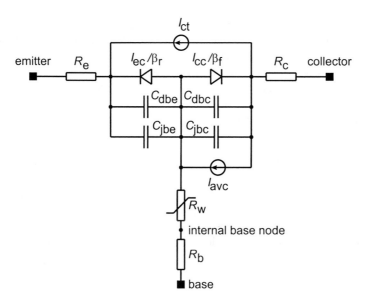

Figure 6.11: EM model extended by junction and diffusion capacitances defining the model behaviour under fast transients.

through the device I_device. They account for the differential resistance of the BJT in the on-state R_diff. Additionally, it should be remembered that β degrades at high injection levels (Blicher, 1981). Wolf (1998) gives an empirical formula for the dependence of β on the collector current

$$\beta = \left(c_\beta (I_\text{c} - I_\text{c0}) + \frac{1}{\beta_0} \right)^{-1} \tag{6.31}$$

with fitting parameters I_c0 and c_β, and the current gain under low injection conditions β_0.

Dynamic behavior of the BJT

In order to reproduce correctly the dynamic behaviour of the BJT, capacitances must be added to the EM model of Figure 6.10 (see Figure 6.11). In analogy to the GP model, the base–emitter and base–junction capacitances can be split into a component accounting for the junction capacitances and a component accounting for the diffusion capacitance. Refining the basic EM model in this way results in increased effort for the parameter extraction however, it is necessary to ensure a correct simulation of fast transients.

The diffusion capacitances C_dbe and C_dce can be modelled by (Massobrio, 1993)

$$C_\text{dbe} = t_\text{f} \frac{\partial I_\text{b}}{\partial V_\text{be}} = t_\text{f} \frac{q}{k_\text{B} T} I_\text{sat,be} \exp \left\{ \frac{q V_\text{be}}{k_\text{B} T} \right\} \tag{6.32}$$

$$C_\text{dbc} = t_\text{r} \frac{\partial I_\text{b}}{\partial V_\text{bc}} = t_\text{f} \frac{q}{k_\text{B} T} I_\text{sat,bc} \exp \left\{ \frac{q V_\text{bc}}{k_\text{B} T} \right\} \tag{6.33}$$

where t_f and t_r are the so-called forward and reverse base-transit times. Hence, the diffusion capacitances define the turn-on time of the BJT, and they account for the diffusion velocity of the electrons in the base region. The base transit times can either be measured directly by vf-TLP experiments (see Section 6.4.3), or they can be estimated. If the BJT is symmetric they are given by

$$t_f = t_r = \frac{L^2}{\eta D_n} \tag{6.34}$$

where L is the base width and η is a coefficient ranging from 2 for low-bias conditions to 4 in the case of high injection (Muller, 1986). For a parasitic bipolar transistor in an NFET with a gate length of 0.2 µm, typical base transit times are in the range of 10 ps (see Section 5.6.2). Such a time certainly does not play a role during HBM stress, but might be considerable for CDM discharges.

The charge stored in the depletion region of the junctions is modelled by the junction capacitances C_{jbe} and C_{jce}. These charges are crucial for the dV/dt turn-on behaviour. The junction capacitances are non-linear functions of the voltage across the junction (Massobrio, 1993)

$$C_{jbe} = \frac{C_{jbe0}}{(1 - V_{be}/\phi_e)^{m_e}} \tag{6.35}$$

$$C_{jbc} = \frac{C_{jbc0}}{(1 - V_{ce}/\phi_m)^{m_c}} \tag{6.36}$$

where C_{jbe0} and C_{jbe0} are the zero-bias junction capacitances, ϕ_e and ϕ_c the built-in potentials and the exponents m_e and m_c are parameters between 0.33 and 0.5, depending on the type of junction. Usually, the zero-bias junction capacitance can be derived from given parameters in the design manual and from the geometry of the BJT.

Extension of the model

Even with the rather sophisticated model defined in the previous paragraphs, which is often used for ESD simulation, there are some shortcomings which, under certain conditions, can lead to severe deviations from the real device behaviour. There are three major simplifications, namely an accurate substrate modelling, resistance modulation and thermal coupling, which are not considered in the basic model.

- In the simplified model discussed above the substrate is modelled as one single resistor. Depending on the layout, this might be too much simplified and a detailed substrate resistance modelling is required, for example to reproduce the trigger behaviour correctly. Li (1998) introduced a substrate resistance network and an extraction methodology for CMOS I/O circuits.

- For all terminal resistors, effectively the same considerations hold as discussed in Section 6.3.2. Because of the high electric field, the electron mobility degrades from its constant low-field value (Puvvada, 2000), leading to a resistance modulation with the voltage across the junction V_j

$$R = R_{\text{low}} \sqrt{1 + \left(\frac{V_j}{V_{\text{crit}}}\right)^2} \qquad (6.37)$$

with the fitting parameter V_{crit}. Equation 6.37 is only valid for low-injection currents dominated by the drift component. According to Boselli (2001), at high injection levels, such as when the BJT is turned on, the diffusion current of the forward-biased emitter–base junction is the dominant contributor to the current across the junction. In that case, the base resistor behaves as $R \propto \exp(\sqrt{I_{\text{be}}})/\sqrt{I_{\text{be}}}$ (Boselli, 2001). Gao (2002a) gives an empirical expression for the substrate resistance $R_{\text{sub}} = R_0 + R_1 \exp(-V_{\text{ds}}) + R_2 V_{\text{ds}} \exp(-V_{\text{ds}})$. R_0, R_1, R_2 are polynomial functions of V_{gs}. In Gao (2002b) an implementation in a SPICE model is presented.

- It is undoubtedly true that self-heating plays an important role in the analysis of devices under ESD stress. During an ESD event the temperature inside a device can reach the melting temperature of silicon. However, it is not only in that extreme region of the IV characteristic of a device that dissipated energy influences the device behaviour. Parameters like reverse saturation currents, the carrier mobility, the multiplication factor M, the ionization coefficient α and consequently the breakdown voltage, as well as the bipolar gain, are at least weak functions of the temperature (Sze, 1981; Muller, 1986). Hence, for a correct analysis, temperature effects must be considered.

As a first step, the heat sources in the device during an ESD stress must be identified. From the device simulation, it is known that at low and moderate current levels most of the power is dissipated in the collector–base junction. Depending on the layout, at higher currents the power dissipated in the terminal resistances R_c and R_e must be considered. Two different basic approaches are known to incorporate the heat source and its influence on the device into the compact model.

- Diaz (1994) implemented a "thermal" integrator in the circuit simulator iETSIM which solves a simplified heat transfer equation (Equation 5.8) in the vicinity of a rectangular box representing the heat source. The "thermal" integrator is analogous to an electrical integrator, using a heat source which converts the electrical power into a temperature change, see (Diaz, 1994; Ramaswamy, 1996).

- In a second approach the dissipated power is modelled as a heat source and a thermal capacitance (Russ, 1993; Luchies, 1994; Russ, 1996; Verhaege, 1996; Russ, 1999). The heating in the vicinity of the heat source is described by a thermal network consisting of discrete thermal resistors and thermal capacitances. The thermal capacitances in this network amount to cylindrical concentric shells around the heat source with logarithmically increased thicknesses from the heat source to the device border.

Alongside these two main approaches, there are further ideas for how the thermal network can be refined (see, for example, Kurimoto 1994). However, it is doubtful whether these relatively complex networks could be used in compact models applied to circuits with more than one device.

Although there have been considerable achievements using one or other of the approaches described above, there are many open issues which make an electro-thermally coupled circuit simulation a rather difficult task. Some of the main issues are:

- The complexity of the compact model increases enormously. The increased complexity has an adverse effect on the numerical stability of the model, making a simulation of even small circuits containing some (2–10) of those models an elaborated, time consuming – and often at the end unsuccessful – task. A further drawback is the significantly higher number of parameters which must be extracted. The lumped elements of the thermal networks, i.e. thermal resistors and capacitances, as well as the heat source, must be determined either from mostly indirect measurements or from rough estimations from layout and/or device simulation.

- The parameters of the thermal network are not very well defined. Thermal resistances and capacitances can only be estimated by layout considerations, or by comparing the final simulation results with experiment. In the second case, a large uncertainty remains because the large number of parameters makes a real fitting procedure almost impossible.

- It is questionable whether electro-thermal circuit simulation without considering 3D aspects is reasonable at all. Often thermal coupling is used to predict the failure current of a device through thermal over-stress. As discussed in Section 5.4.5, even for 3D device simulation this is only possible if the simulation doping profile is accurate and reproduces the "real" device. Circuit simulation is more a 1D representation of the device, so effects such as current confinement, filamentation, and successive melting of the silicon are not described. If a "quasi-3D" network is defined for the compact model, again the numerical behaviour is fatally deteriorated, and a better choice is to use a device simulator.

- The temperature dependence of the parameters used in the models is valid at the utmost for $T < 800$ K. In the case of an ESD event the temperature can reach the melting point of silicon (1685 K), far above the scope of application of the present model. Therefore, circuit simulation in that temperature range carries a significant uncertainty, and appropriate estimates have to be made.

There is no single answer to the question of whether thermal effects should be considered in the compact models. Depending on the problem to be solved, one has to decide whether thermal coupling must be included, always bearing in mind that by doing so, the complexity of the compact model increases significantly, and that severe convergence problems can therefore be expected. This makes simulation

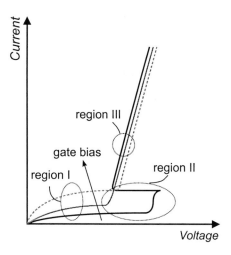

Figure 6.12: Schematic representation of the *IV* characteristic of an NFET with different
gate bias. Region I: linear and saturation characteristic, region II: breakdown
and snapback characteristic, region III: high-current characteristic.

of large circuits absolutely impossible (see examples in Section 6.5). Having said
this, there are areas, e.g. SOI technologies, where neglecting thermal coupling
leads to unrealistic results due to the increased temperature, in comparison to
bulk material, even at lower currents (Raha, 1997).

6.4 Compact models of FETs

6.4.1 General considerations

The behaviour of FETs under standard operating conditions, as well as under
ESD stress, has been widely discussed and is very well known (see, for example,
Amerasekera, 2002). In Figure 6.12 schematic *IV* characteristics for the NFET
with different gate bias are shown. The *IV* curve consists essentially of three
regimes: first the standard operating regime (I) covering the linear and saturation
region of the MOSFET; second, the breakdown regime with avalanche breakdown
and snapback (II); and, third, the high-current characteristic after snapback (III).
The characteristic in region I is exclusively determined by the NFET behaviour.
The breakdown and snapback in region II is influenced both by the MOSFET and
the parasitic *npn* bipolar transistor, while region III is mainly governed by the
parasitic *npn* bipolar transistor, with only the holding voltage being influenced by
the gate potential (Hsu, 1982)

A lot of excellent work has been done on the circuit simulation of FETs. The
early work investigated models for the grounded-gate NFETs (Russ, 1993; Luchies,
1994; Russ, 1996). These studies focused on the behaviour of the parasitic bipolar transistor, more or less neglecting the FET; this is a sound approach for the
grounded-gate configuration where the FET is switched off. The parasitic bipolar

model in all the studies was based on the EM model discussed in detail in Section 6.3.3. Amerasekera (1996) developed a SPICE® (Synopsys Inc., Mountain View, CA) model equation set for the behaviour of the NFET under ESD stress, considering both the FET and the parasitic bipolar transistor. Based on this work, Ramaswamy (1996) introduced a BSIM3 FET model that is extended by the parasitic bipolar action, including transient effects and gate-coupling as well as thermal coupling. Using the circuit simulator iETSIM, the model yields excellent results in terms of reproducibility of experimental data of NFETs under electrical overstress and CDM stress. The problem of this approach is, however, that the MOSFET parameters used in this model are not necessarily identical to the MOSFET parameters of the proprietary models of the semiconductor manufacturers which are used for circuit simulation in the standard operating regime. Therefore, combining the standard MOSFET models and the ESD MOSFET model might be difficult. To overcome this drawback, the most common solution nowadays is to use a modular approach, in which the parasitic bipolar transistor and the other ESD-relevant elements are combined with the standard BSIM MOSFET model, without changing the standard BSIM model. Examples of such a strategy are given in Lim (1997), Wolf (1998), Beebe (1998), Mergens (1999a), Mergens (2000), Wolf (2001). The advantage is obvious. The ESD engineer can use the standard models for the FET, including the extracted standard parameters that are, in any case, inevitable for the functional simulation of circuits in the standard operating regime. Therefore, the ESD engineer need not bother with the time-consuming and complicated parameter extraction procedure for the standard MOSFET model.

The use of a modular compact model consisting of at least a standard MOSFET model and a compact model for the parasitic bipolar transistor will be explained in detail in the following section. The parameter extraction procedure and the verification of the model, together with its limitations, will be discussed in Section 6.4.3.

6.4.2   The modular FET compact model

Figure 6.13 shows one possible approach for a modular compact model of an NFET. The following discussion will focus on the NFET; the PFET is closely analogous.

Basically, the model consists of an NFET model which accounts only for the FET operating regime, a BJT compact model dealing with high-current effects, an n^+p diode from drain to bulk accounting for ESD stress in which the drain–bulk diode is forward biased, and finally a sub-circuit modelling the source–bulk diode. The four main parts are marked in Figure 6.13 by dashed boxes.

For the MOSFET compact model a standard BSIM model (usually either a BSIM3 or BSIM4 model[2]) is used. It is worth repeating that the model itself is not modified in any way. Detailed descriptions of BSIM models, which are beyond

---

[2]BSIM3 and BSIM4 are developed by the Device Research Group of the Department of of Electrical Engineering and Computer Science, University of California, Berkeley and copyrighted by the University of California. The latest implementations of the BSIM models are available at http://www-device.eecs.berkeley.edu

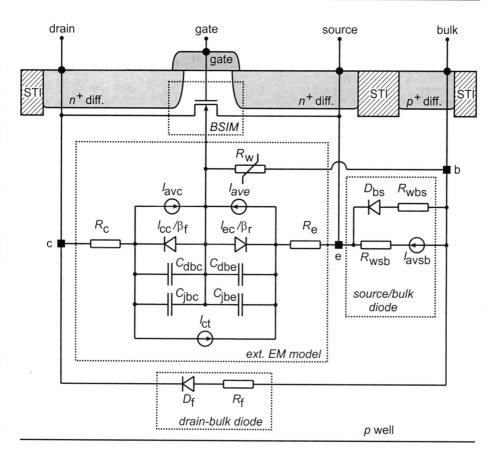

Figure 6.13: Modular compact model for an NFET. The compact model consists of a BSIM MOS model, a BJT based on an EM model (with R_b included in R_w, a diode accounting for the drain– bulk diode under forward biased conditions, and a avalanche source between source and bulk.

the scope of this book, can be found in, for example, Massobrio (1993). The ESD engineer does not need a detailed understanding of the BSIM model. However, as we will see in Section 6.4.3, some parameters might have to be changed with respect to the "stand-alone application" of BSIM models for circuit simulation in the standard operating regime.

The compact model for the parasitic BJT was discussed in detail in Section 6.3.3. However, two additional features have been implemented in comparison to the model presented in Figure 6.11. First, a coupling of the FET and BJT must be considered in the model. Secondly, in general an avalanche source I_{ave} is included to account for the avalanche breakdown of the base-source junction.

Clearly, the BSIM model does not only define the standard operating regime (region I in Figure 6.12), but it also influences the breakdown and triggering behaviour of the FET (region II in Figure 6.12). The main reason for this is the additional MOS current I_d which depends on the gate–source voltage V_{gs}. This MOS current adds to the avalanche current I_{avc}. Consequently, Equation 6.22 must be extended to

$$I_{avc} = (M - 1) \times [I_{ct} + I_d(V_{gs})] \qquad (6.38)$$

The basic equations of the BJT can be deduced straightforwardly from the corresponding equations in Section 6.3.3. Note, however, that in general, owing to the parasitic BJT, the current distribution in the device may be completely different for standard operation and under ESD stress. Consequently, the parameters of a BJT may have a very different meaning in the case of a parasitic lateral bipolar transistor (see, for example, the discussion about current gain β in Amerasekera (1996)).

Just as with the avalanche current source I_{avc}, a second avalanche source I_{ave} has to be included. This source is necessary to describe ESD stress where the source potential is higher than the drain potential. Under such stress conditions, the emitter–base junction of the parasitic BJT is driven into reverse mode and the collector–base junction is forward biased. Consequently, avalanche generation can start at the emitter–base junction, analogous to the conditions discussed in Section 6.3.3 for the collector–base junction. Such an extension of the basic model is particularly important for CDM stress where, in general, stress of both polarities usually occurs during a discharge (Russ, 1996; Beebe, 1998; Mergens, 1999).

If the bulk and source are not connected together, a voltage drop can occur between these both terminals. Therefore, the model has to include an additional avalanche source I_{avsb} with a series resistor R_{wsb} accounting for a possible avalanche breakdown between source and bulk. A diode D_{bs} and a series resistance R_{wbs} in parallel model positive stress at the bulk. These additional features make the model and the parameter extraction procedure more complex; however, this is inevitable in order to handle substrate separation correctly.

In general, the modular compact model presented in Figure 6.13 is rather user-friendly in terms of parameter extraction procedure, numerical stability, and run-time behaviour. However, depending on the problem to be analyzed, the model could be either reduced (see examples in Section 6.5) or else further elements and physical models may be required. A typical example is electro-thermal coupling. Increased temperature owing to power dissipation will also have an influence on the MOS models; for example, the drain–source current I_{ds} will be decreased. It is safe to assume that the temperature dependence of the MOS parameters is implemented in the BSIM model in a very sophisticated manner; there is no need to extend the models towards even higher temperatures because the MOS has no influence on the high-current characteristic. The turn-on behaviour alone might be slightly changed by the thermal coupling. If needed, thermal coupling can be included in the BJT model as outlined in Section 6.3.3. However, the numerical stability and the run-time behaviour will be degraded dramatically, and the effort required for parameter extraction procedure will be significantly increased.

6.4.3   Parameter extraction and verification of the model

The parameters used in the modular compact model presented in the previous section can be classified into four groups according to the four main modules of the compact model.

For the ESD engineer, the parameters that can most easily be obtained are the FET parameters, (often parameters for the BSIM3 or a BSIM4 model). Parameter extraction may be conducted by a special device modelling group, whose function is to provide the model, including its parameters, to all circuit designers. The model parameters are in general available at an early stage of the process development, because they are the basis for all circuit design. The ESD engineer may not need to know the BSIM model and its included parameters in detail, but he/she must make sure that parameters treated in the BSIM model and the ESD-specific add-on are not double counted. Typical candidates for being cancelled or set to zero are the junction capacitances and parameters which model the breakdown in the BSIM model, i.e. the substrate current parameters.

The extraction of parameters for an ESD specific add-on must be done separately. A comprehensive list of the key parameters and the corresponding extraction methodology is given in Table 6.1. A cautionary warning is necessary. The parameters depend strongly on the layout and on the measurement conditions used for the parameter extraction. As an example, consider the bulk contact configuration and the bulk potential during the extraction; these could have a significant influence on the region where the breakdown starts, on the turn-on current, and on β.

Table 6.1: Compilation of the most important parameters for the NFET compact model (PFET in analogy). The MOSFET parameters included in a BSIM model are neglected.

Parameter	Meaning	Extraction
parameters of extended EM model		
M_r	reverse multiplication factor for I_{avc}	estimated from DC measurements in the avalanche region before snapback; $M = I_d/(I_d - I_{bulk})$
M_f	forward multiplication factor for I_{ave}	estimated from DC measurements in the avalanche region before snapback; $M = I_s/(I_s - I_{bulk})$ or assumed $M_r = M_f = M$
β_f	current gain for positive stress on drain	TLP measurements with monitoring drain current I_d and bulk current I_{bulk}; $\beta_f = I_d/(I_d \times (M_r - 1) - I_{bulk}) \times M_r$ or fitting on holding voltage

Table 6.1: continued

Parameter	Meaning	Extraction
β_r	current gain for positive stress on source	TLP measurements with monitoring drain current I_s and bulk current I_{bulk}; $\beta_r = I_s/(I_s \times (M_f - 1) - I_{bulk}) \times M_f$ or $\beta_f = \beta_r = \beta$ or fitting on holding voltage
R_c	collector resistance	TLP measurements in reverse mode and extraction of differential resistance for varying layout parameter DCG, alternatively: extraction from layout
R_e	emitter resistance	TLP measurements in reverse mode and extraction of differential resistance for varying layout parameter SCG, alternatively: extraction from layout
R_w	well resistance in the low-injection regime	DC measurements in reverse mode
C_{jbc0}, C_{jbe0}	zero-bias junction capacitances	extraction from layout, alternatively, dV/dt triggering and fit
C_{dbc}, C_{dbe}	diffusion capacitances	determine base transit time either by fitting vf-TLP measurements or by calculation $t_f = t_r = L^2/(\eta D)$, then using Equation 6.32 and 6.33; see Section 6.3.3
parameters of drain–bulk diode		
R_f	serial resistor for drain–bulk diode	extraction of differential resistance from TLP measurements in forward mode
n	ideality factor of drain–bulk diode	extraction from TLP measurements in forward mode
I_s	reverse saturation current of drain–bulk diode	extraction from TLP measurements in forward mode, alternatively, estimation from layout, low-current DC measurements,

Table 6.1: continued

Parameter	Meaning	Extraction
parameters of source–bulk avalanche source		
M_{r}	reverse multiplication factor for I_{avsb}	fitting from source–bulk TLP measurements in reverse (avalanche) mode
R_{wsb}	series well resistance to avalanche source between base and source	fitting from source–bulk TLP measurements in reverse (avalanche) mode
n	ideality factor of source–bulk diode	extraction from source–bulk TLP measurements in forward biased mode
R_{wbs}	series well resistance to source–bulk diode between base and source	extraction from source–bulk TLP measurements in forward biased mode

In the interests of a consistent ESD simulation flow, it is important to mention that most of the parameters can be obtained by ESD device simulation. The multiplication factors and the current gains can be calculated from the corresponding currents, which are a direct outcome of the device simulation. The resistance of the well and the collector can be derived from the 2D simulation of the high-current characteristic. Finally, the base transit time can be calculated directly in the device simulation (ISE–TCAD, 1998). Furthermore, the DC characteristic (Figure 6.14), the TLP high-current characteristics (Figure 6.15) and the response of the device to a vf-TLP stress which are used as verification of the model and its parameters (see discussion in the following paragraphs), can be obtained by device simulation, too. Therefore, device simulation cannot only be used for the optimization of single devices, but helps also to design ESD-optimized circuits by providing parameters for the compact models used in the ESD circuit simulation. This is a further strong benefit for a consistent ESD simulation flow.

Having completed the parameter extraction, the model should be verified in the ESD-relevant and MOS regimes. In the following, some typical results for an NFET in a 0.18 µm CMOS technology are discussed.

First, the FET standard operating regime should be analysed. Although, at first glance, the FET range seems to be secondary for an analysis of ESD problems, this is not the case if large circuit blocks are investigated. In such cases, the FET behaviour can indeed have an influence on the ESD robustness where, for example, an FET in a pre-driver logic controls the gate potential of a driver in the discharge path. Using the BSIM model alone, good agreement between circuit simulation and experiment in the whole specified operating regime must be assumed. However, the circuits parallel to the BSIM MOSFET model, in particular the BJT model, can always modify the results. Figure 6.14 shows the DC characterization of an NFET with different gate bias, first with an FET and a BJT in parallel and then with the FET as a stand-alone device. Up to a gate voltage of 1.6 V, the

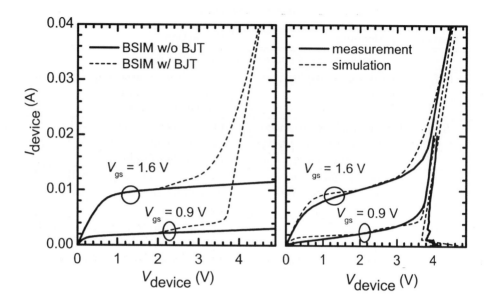

Figure 6.14: DC characteristic of the NFET. The bipolar model extends the MOS model
          towards the high-current domain without having a significant influence on
          the standard MOS operation.

circuit simulation with both set-ups shows a reasonably good correlation with the
experimentally obtained IV characteristics. In this voltage regime, no difference
can be seen for the FET model and the compact model with a BJT parallel to
the FET. From this, one can conclude that in the operating regime of the NFET
($V_{\text{device}} \leq 1.8$ V), the parallel circuit of the BJT and the drain–bulk diode does
not influence the FET mode.

The high-current characteristics obtained by TLP for the most important stress
configurations are summarized in Figure 6.15. The good agreement between simulation and experiment for the high-current behaviour of the source–bulk junctions
(Figure 6.15 a)) verifies the R_{wsb} and I_{avsb}. Similarly, the TLP characteristic of
the bulk–drain junction under forward bias (Figure 6.15 b)), which is defined by
the elements D_{f} and R_{f}, is reproduced quite well by circuit simulation. In Figure 6.15 c), the high-current characteristic for reverse-bias stress of the NFET is
depicted for a grounded-gate configuration. Good correlation can be obtained for
low current densities. However, self-heating effects can be observed, which cause
an additional voltage drop for higher current densities under 100 ns TLP stress.
It is worth to mention that a crucial issue for the usability of a compact model
in an ESD circuit simulation is its scalability. This is particularly important for
FETs which might exist in an I/O circuit with several different device width. From
Figure 6.15 a)–c) one can conclude that scalability is given as long as self-heating
effects are negligible.

That the deviation of the simulation from the experimental data is indeed

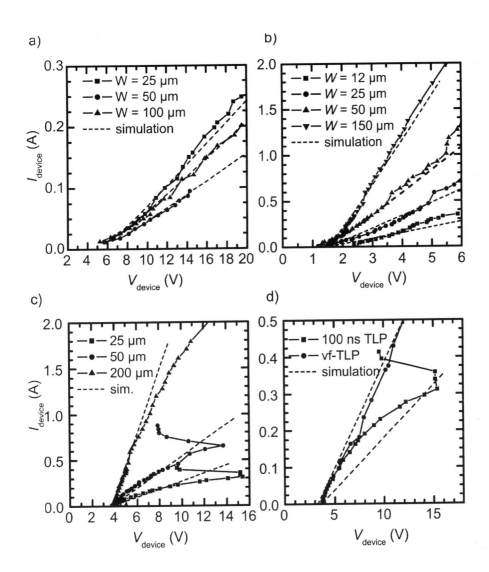

Figure 6.15: TLP characteristics of an NFET with several different stress configurations.
a) Source–bulk junction in reverse mode for different gate width.
b) Bulk–drain junction under forward-biased conditions for different gate width.
c) NFET in reverse mode for different gate width.
d) NFET in reverse mode under 10 ns vf-TLP and 100 ns TLP stress.

due to self-heating effects is demonstrated in Figure 6.15 d), where results obtained by 100 ns TLP are compared to those obtained by 10 ns vf-TLP. For 10 ns vf-TLP, self-heating effects are negligible, and consequently, circuit simulation without electro-thermal coupling and vf-TLP experiments match very well. For longer pulses and smaller devices, which show higher energy dissipation per volume, the correlation worsens. In order to take into account self-heating effects in the compact model, a thermal network must be included. Successful calibration of a thermal network by means of isothermal TLP characterization at elevated temperatures and transient measurements is described in Wolf (1996). However, in the light of the major drawbacks (numerical stability and the parameter extraction) discussed in Section 6.3.3, an alternative approach is proposed in the following which will be used for the simulation of complete I/O cells in Section 6.5.

To estimate the effect of the self-heating of the protection device, a parameter set is defined with an increased differential resistance. The differential resistance which includes self heating effects is extracted from a high-current characteristic of a small device with 100 ns pulses (Figure 6.15 d) which reflects a kind of worst-case situation concerning self-heating during an ESD stress. The circuit simulation has to be performed twice, first with the parameter set in which R_{diff} corresponds to the vf-TLP experiments (without considering thermal effects), and second with the parameter set corresponding to the 100 ns TLP stress (worst-case conditions). From these both simulation runs, the minimum and maximum voltages can be extracted. Certainly, this procedure has the disadvantage that there is only a worst-case estimation which not necessarily reflect the real behaviour of the device. However, the robustness and applicability of the compact model is significantly improved, which is particularly important for the simulation of large circuits.

Finally, the transient (turn-on) behaviour of the compact model must be verified. For this verification, a method is required which can resolve the device behaviour during the first few tens of picoseconds. The only experimental method which has been successfully applied to this problem is vf-TLP, as introduced by Gieser (1996). Figure 6.16 shows the response of the device on a 10 V pulse for two different gate widths. For more detailed information about the analysis of vf-TLP experiments see Wolf (2000). According to Equation 6.34, the base transit time depends on the base width which correlates in the case of the NFET with the gate width. Experiment and simulation show very good agreement, indicating that the diffusion capacitance parameters in the model are correctly defined.

6.5 Examples: circuit simulation of ESD problems

In this section, some examples will be given of how to apply the compact models, which have been discussed in the previous section, to "real" ESD problems. The objective of the examples is clearly to highlight the power of the ESD circuit simulation, and also to indicate the limitations of the current models and the whole approach of simulating ESD problems by means of circuit simulation. All the examples have been investigated in deep sub-μm technologies with feature

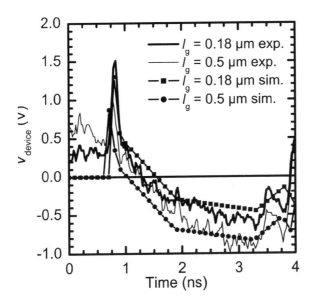

Figure 6.16: Vf-TLP experiments on a ggNFET for two different gate lengths. Only the reflected pulse is drawn; this represents the response of the device to the incident pulse (Wolf, 2000).

size 0.18 µm and 0.13 µm. For the circuit simulation, SABER has been used for the 0.18 µm process, while for the 0.13 µm technology VHDL–AMS models have been used in a proprietary simulator (TITAN). For the examples, several blocks of I/O cells, together with the corresponding power supply cells, have been included. The I/O cells also contain intentionally ESD critical designs and layouts. The calculated circuits involve 5–70 single elements, exclusively NFETs and PFETs, diodes, and resistors as discussed in Sections 6.3 and 6.4.

As a first step, the netlist of the I/O stages or circuits under consideration has to be modified in order to introduce the ESD-specific compact models. Here two approaches have been used:

- All devices have been replaced by their corresponding ESD compact models (all elements are simulated correctly in the high-current regime).

- Only those devices which are either connected directly to the pad or have a layout with ESD measures (silicide blocking, enlarged DCG) are replaced by their corresponding ESD compact models.

The runtimes of HBM circuit simulations using both approaches are compiled in Table 6.2 for three different I/O cell libraries, each consisting of 20–70 devices. Additional runs have been performed including thermal coupling. It is clear that neglecting the thermal coupling significantly improves the run-time and convergence behaviour of the compact models. Experience with SABER simulations shows that an appropriate strategy could be first to replace only those elements

Table 6.2: Run-time behaviour of different HBM simulations depending on the simulation
strategy and the model (thermal coupling). For the simulations a Sun Ultra 20,
512 MB memory, was used.

Approach	Thermal coupling	runtime I/O cell library		
		1.8 V	3.3 V	2.5 V / 3.3 V tol.
Approach 1: all elements are replaced by ESD compact models	no	no convergence	~ 25 min	~ 30 min
	yes	no convergence	no convergence	no convergence
Approach 2: only elements connected to a pad are replaced by ESD compact models	no	~ 40 s	~ 45 s	~ 40 s
	yes	~ 11 min	~ 12 min	~ 11 min

with ESD layout measures (e.g., elements connected directly to the pads) by the
appropriate ESD compact models. Based on the results of the first simulation
run, devices which carry a significantly higher current during an ESD event than
during normal operation, or are connected to nodes which significantly exceed the
operational regime, should also be replaced by their corresponding ESD compact
models. The simulation is then repeated. This approach can save significant run-
time in comparison with the "straightforward" approach in which all elements are
replaced by ESD compact models. Often, the "straightforward" approach fails on
account of convergence problems.

Before starting the ESD circuit simulation, appropriate failure criteria have
to be defined. As discussed in Section 5.4.5, in the ESD device simulation internal device parameters as maximum temperature and maximum electric field
strength across the gate oxides are monitored, as well as the occurrence of second
breakdown in the case of 2D device simulation, see Equation 5.18. However, these
parameters are only hardly (temperature) or not at all (electric field across gate
oxide, occurrence of second breakdown) accessible by ESD circuit simulation. For
ESD circuit simulation a straight-forward approach is the definition of maximum
current densities and maximum voltages at circuits nodes

$$I_{t2,sim}/W \equiv \min\{ \quad J > J_{max} \text{ for all devices of the circuit,}$$
$$J @ (V > V_{crit}) \text{ for all nodes in the circuit } \} \quad (6.39)$$

The devices in which the current density during an ESD stress exceeds a pre-
defined limit, say, the maximum current density obtained by device simulation,
and circuit nodes in which the voltage exceeds the ESD design window must be
flagged by the circuit simulator.

a)   b)

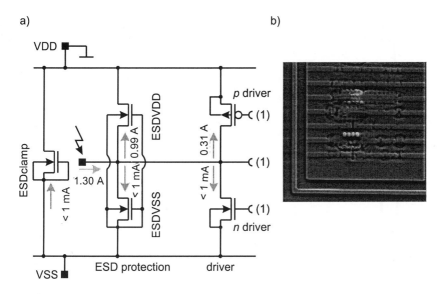

Figure 6.17: a) Output cell with ESD relevant elements considered in the simulation. The currents for a 1.3 A transmission line pulse in the steady state (at about $t = 50$ ns) are included. The whole current is shunted across the protection element to VDD (ESDVDD) and the p driver. The nodes (1) are connected to the pad control logic. b) The physical failure analysis shows molten diffusion areas across the whole p driver caused by a non-ESD optimized layout.

6.5.1 Output stage

In the first example, a failure in an output cell is verified by circuit simulation. The schematic of the I/O circuit reduced to its ESD relevant elements is depicted in Figure 6.17 together with the result of the physical failure analysis which clearly shows several molten filaments along the gate of the PFET. The root cause for this damage is a non-ESD-optimized PFET layout, which degrades the ESD robustness of the PFET from > 4 mA/µm (ESD optimized layout) to ~ 1.0 mA/µm.

The steady-state current distribution, i.e. the current distribution at the HBM peak current for an HBM discharge voltage of $+2$ kV and a 1.3 A TLP (risetime 1 ns) in the output pad, is drawn in Figure 6.18. The input nodes of the pad logic do not play any significant role for the current distribution; however, in order to avoid convergence problems these pins are tied to ground via a capacitor. The current in the shunt path "pad $\rightarrow$ ESD protection element ESDVSS $\rightarrow$ VSS rail $\rightarrow$ ESDclamp $\rightarrow$ VDD" can be neglected. In contrast, the discharge path "pad $\rightarrow$ p driver $\rightarrow$ VDD rail" plays an important role. Although the current through the PFET is only one third of the current through the protection element, the current exceeds the PFET's maximum withstand current. The device width of the PFET is $W = 235$ µm, resulting in a maximum withstand current of $I_{\mathrm{HBM}} = 1.0$ mA/µm $\times$ 235 µm ≈ 235 mA for the non-optimized ESD layout of the PFET. A stress of 2.0 kV causes a current of 310 mA through the PFET, which causes

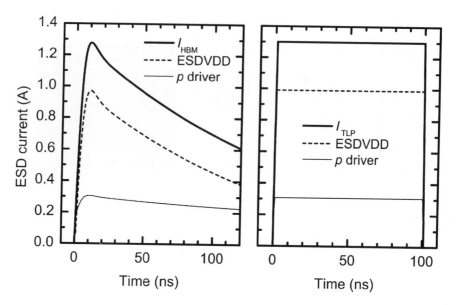

Figure 6.18: Current distributions in the output cell of Figure 6.17 for a 2 kV HBM
discharge (left) and a 1.3 A transmission line pulse (right).

an over-stress. This result corresponds to the findings of HBM tests, in which this
pad fails at about 2 kV, passing 1.5 kV.

There are several layout and design possibilities for optimizing the ESD robustness of this particular I/O cell. Clearly, the simplest way would be to increase the
intrinsic ESD robustness of the PFET to 5–10 mA/µm which is usually obtained
for PFETs. However, there might be reasons why the intrinsic ESD robustness of
the PFET cannot be increased – technology measures may be impossible because
of a frozen technology, and performance reasons may prevent layout measures. In
such cases, increasing the width of the ESD protection element to VDD can reduce
the stress to the PFET.

At first glance, the positive discharge of the pad to VSS seems to be similarly
easy. The current distribution is expected to behave according to the width of
the NFETs, i.e. the n driver and the ESD protection element ESDVSS. Indeed,
in the steady state of a TLP discharge, the ratio of the TLP current through the
devices, ESD and n driver ≈ 0.875 A/0.425 A, corresponds to the width ratio of
200 µm/97 µm (see Figure 6.19). However, in the first nanosecond, i.e. during the
ramp-up of the pulse, a current through the p driver to VDD can be observed as
soon as the voltage at the pad exceeds the on-voltage of the parasitic p^+n diode of
the p driver. The reason for the current to VDD is that the capacitance between
VDD and VSS is initially charged up through the forward-biased parasitic p^+n
diode of the p driver. At the end of the transmission line pulse, the voltage at VDD
remains at about 7 V. In the simplified example, the charge stored in the chip

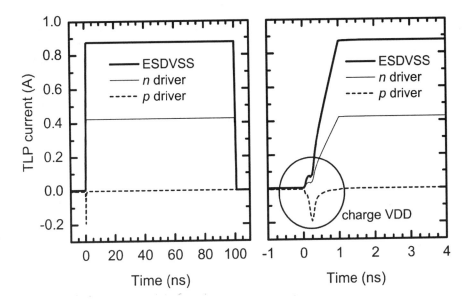

Figure 6.19: Positive 1.3 A square pulse discharge to VSS. As expected, in the steady-
state regime the current is shunted to VSS via the ESD protection element
ESDVSS and the n driver (left), the portion being proportional to the width
of the devices. During the first nanosecond, the current flows through the p
driver and charges the chip capacitance between VDD and VSS (right).

capacitance cannot be balanced, because the voltage at VDD is smaller than the
breakdown voltage of the p driver and the ESD protection element between pad
and VDD. In reality, the capacitance is discharged slowly by the leakage current
between VDD and VSS, an effect which could be included by a high-ohmic resistor
between VDD and VSS.

Obviously, the peak current and the charge flowing through the PFET will
depend on the total capacitance of the IC (see Figure 6.20). As soon as the
chip capacitance exceeds the built-in capacitance of the power clamp protection
elements, the charge driven through the diode Q depends linearly on the capac-
itance, $Q = C_{\text{chip}} \times V$, where C_{chip} is the chip capacitance and V is the voltage
under steady-state conditions. The peak current increases with increasing C_{chip},
until the whole discharge current is driven through the p driver and the current is
limited by the TLP current (1.3 A in this example).

Although the analysis in the above example is rather simplistic, it strikingly
shows the problem of choosing appropriate lumped elements for the IC. Where
guidelines are derived from the circuit simulation, the definition of correct boundary conditions, i.e. the parasitic elements of the circuit which reflect the worst-case
conditions, is a crucial task of the ESD engineer. In the example of Figure 6.19
a current through the p driver defined by the capacitance of the power clamps
(without additional chip capacitance) might be not critical. However, it must be

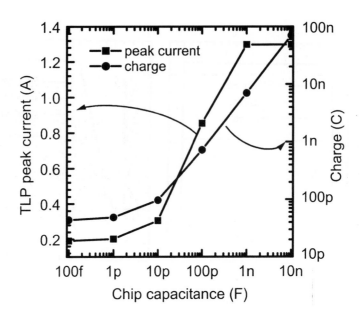

Figure 6.20: Dependence of the peak current (left axis) and the total charge (right axis)
through the p driver in the case of a 1.3 A TLP current.

borne in mind that under worst-case conditions the whole current could be shunted
via the p driver, if the chip capacitance is large enough. Actually, a large chip capacitance can be very beneficial for ESD protection. For a positive ESD stress at
the I/O pad and VSS grounded, the current is shunted to the capacitance via the
forward biased diode, which in general is rather ESD robust. The discharge occurs
via the leakage paths, without noticeably stressing the elements to VSS.

The results of these simulation studies are not surprising: similar results could
be obtained by simply estimating the current distributions between NFETs and
PFETs from their high-current characteristics and calculating the charge which
can be stored in a capacitor. Such estimations can be done by hand for simple
If the circuitry becomes more complex, an automated simulation is necessary to
handle the data and to ensure that no critical path is missed.

6.5.2   Inverter failures in the core

Core fails are much more severe challenges for an ESD circuit simulation. Obviously, it is impossible to perform an analogue simulation of the whole IC with
several millions of transistors. The main task is therefore to reduce the problem
to the ESD relevant elements.

Failures in inverter-type structures in the core of an IC are well known in
the ESD literature (Amerasekera, 2002). However, most of the failures discussed
in literature are the result of the turn-on of parasitic elements, such as a $pnpn$
SCR-type structure which is formed by the neighbouring NFET and PFET and

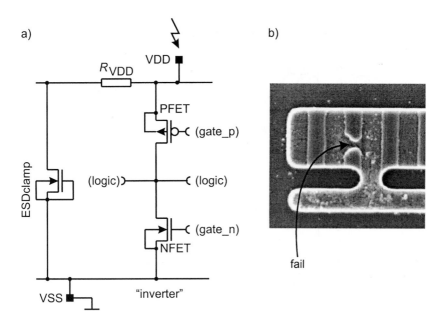

Figure 6.21: a) Inverter-type structure which fails in a test circuit. The ESD analysis
considers the power clamp and a possible bus resistor R_{VDD}. The potentials
at the gate nodes gate_p and gate_n have an influence on the current distribution of the sub-circuit during an ESD event. The connections labelled
"logic" belong to the logic of the I/O cell and do not have a direct connection
to an I/O pad. b) The physical failure analysis yields a melting in one finger
of the NFET of the inverter-type structure.

possibly the surrounding guard rings. Typical failure signatures include melting
between the neighboured diffusions of NFET, PFET and guard rings. In contrast,
in the case study discussed below, melting occurred only at the gate area of the
drain diffusion of the NFET (see Figure 6.21 b)). NFET and PFET are locally far
away from each other, and therefore no parasitic SCR or bipolar can be formed.
Obviously, the root cause for the damage is a thermal over-stress caused by an
excessive current flowing through the inverter-type structure. This phenomenon
itself can be reproduced qualitatively without problems (see Section 2.5.3). During
an ESD event the voltage across the PFET significantly exceeds the voltage during
operation. Depending on the potential at the gate of the PFET, the PFET drives
a high current, which can be large enough to damage the NFET of the core inverter
stage, which has no ESD protection measures. However, the quantitative analysis
is quite complex and very time-consuming. The failure provoked by an insufficient
power clamp and a special configuration of the PFET and NFET in the inverter-
type structure of the logic of an I/O cell is shown in Figure 6.21, together with
the ESD-relevant circuit elements which are used for the analysis of the problem
by means of circuit simulation.

The "fatal" snapback of the NFET in the inverter stage, which leads imme-

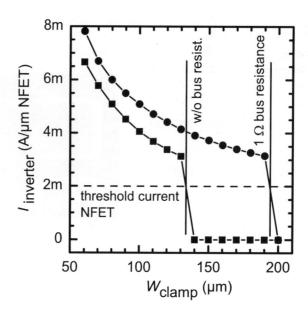

Figure 6.22: Dependence of the current through the inverter I_{inverter} on the voltage clamping capability, i.e. on the width of the power clamp element (here: ggnFET)
and a bus resistance. The power clamp width must exceed 140 μm without
bus resistance and 200 μm for a 1 Ω bus resistance to protect an inverter
structure with the particular width of the PFET and a minimum finger width
of the NFET of 1 μm.

diately to a thermal over-stress for non-ESD optimized layouts (as is mostly the
case for inverter stages not connected directly to an I/O pad), depends mainly on
the voltage drop across the inverter stage. Figure 6.22 shows the current through
the inverter as a function of the width of the power clamp in Figure 6.21 and
the bus resistance. To a first approximation, the potentials of the gates are both
tied to VSS. This configuration should reflect a "worst-case" condition, because
for these potentials the current capability of the NFET in breakdown is worst
and the PFET is completely open, leading to a maximum current. For the analysis, a square pulse was applied to the supply and the voltages and currents were
evaluated in the steady-state regime, i.e. in the region of approximately constant
currents and voltages of the pulse. As long as the trigger condition of the small
NFET is not fulfilled, only a negligible current flows across the inverter. Neglecting
transient effects, the NFET can be turned on when the voltage across the inverter
exceeds the trigger voltage of the NFET plus the threshold voltage of the PFET.
The voltage drop across the inverter is determined by the voltage clamping capability of the power clamp, in other words the type of the power clamp device and
its width, and the parasitic resistance in the VDD and VSS power lines over which
the ESD current is shunted. Assuming an intrinsic ESD robustness of 2 mA/μm
for the non-optimized core NFET device, the critical width of the power clamp can

be calculated. Clearly, even a small bus resistance can have a devastating effect on the inverter current. An more sophisticated power supply concept can help to avoid such fails. In this example, the power clamp width must exceed 140 µm with negligible bus resistance and 200 µm for a 1 Ω bus resistance to protect an inverter structure with the particular width of the PFET and a minimum finger width of the NFET of 1 µm. For this estimation it is assumed that, owing to the missing ballasting resistance of the core NFET, a multi-finger triggering is not guaranteed, and hence that the smallest finger of the NFET determines the ESD robustness.

Efficient design measures to avoid such inverter fails are obviously all measures which minimizes the voltage drop in the power net, either an improved clamping capability of the power supply protection or a reduced bus resistance. Further counter measures could be to limit the current across the inverter-type stage by introducing resistors in series to the inverter or an ESD-optimized layout of the NFET devices in critical voltage domains.

For this simulation, use of ESD compact models for the NFET and PFET is mandatory, or else the breakdown behaviour of the NFET would not be reproduced correctly, resulting in wrong values for the voltage drop across the PFET and, hence, to completely wrong results. However, the simulation is based on several assumptions, which make the correlation with real-world effects harder than it might seem.

1. The potential of the gate of the PFET (gate_p) is defined by the circuit (perhaps an inverter chain) which is connected to the gate. In principle, this circuit has to be simulated completely to extract the correct gate potential. For this particular example it could be shown that the potential gate_p is mainly defined by one certain input pin of the I/O cell. Certainly, the potential at the input pins connected to the core is unknown. Therefore, both high and low potentials have to be investigated. In general, such studies have to be performed for all possible signal combinations at the different pins of an I/O cell! The dependence of the gate potential on the circuit can be seen in Figure 6.23 where the gate potential is defined by lumped capacitances between VDD, gate_p, and VSS which acts as a potential divider. Depending on the capacitances, the current through the inverter is either negligible or critical. Additionally, the potential at the gate of the NFET (gate_n) influences the ESD robustness of the NFET.

2. The ESD withstand current depends, in first order, only on the layout of the NFET and on possible ESD measures. The layout information is not available in the netlist of the pad, making the correct "flagging" of an over-current difficult.

3. The parasitic bus resistances R_{VDD} and R_{VSS} are in general not considered in the schematic of an I/O cell. 1 Ω additional bus resistance causes an additional voltage drop of 1.3 V at 2 kV HBM stress. The net lists have to be modified manually in order to include the parasitic bus resistances. An alternative is the extraction of a back-annotated net list.

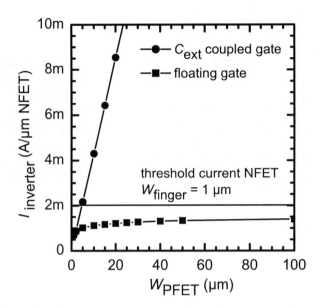

Figure 6.23: Dependence of the current through the inverter on the PFET width and the potential at the PFET gate node. In the case of floating gates the potential at the PFET gate is tied to VDD by the intrinsic capacitances of the PFET, and the current through the PFET is consequently limited. If external capacitances C_{ext} in the range of several tens of fF are used as lumped elements to simulate the circuit connected to gate_p, these capacitances govern the potential at the gate, and the saturation current of the PFET increases linearly with its width. Assuming a finger width of 1 µm and an intrinsic NFET ESD robustness of 2 mA/µm, the NFET will survive a 2 kV stress even for a PFET width of 100 µm for floating gates. With the lumped capacitances the NFET fails even at very small PFET widths ($W_{\text{PFET}} \approx 7$ µm).

4. The maximum voltage across the inverter may be given by the trigger voltage V_{t1} of the protection element at low current levels, instead of the clamping voltage at high current levels. Turn-on is a transient phenomenon of the first nanoseconds of the pulse. This might actually cause the highest current through the inverter stage.

The phenomenon of inverter failures can be quantitatively explained by ESD circuit simulation with the compact models. However, this example also shows the limitations of the circuit simulation. A *predictive* quantitative analysis is limited by the aforementioned uncertainties, which exist in circuit simulation in general.

6.5.3   Simulation of over-voltage tolerant pads

As shown in Table 6.2, the simulation even of comparable small circuits often leads to convergence problems if all elements are replaced by the corresponding model

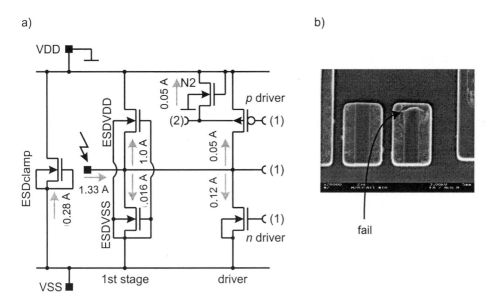

Figure 6.24: a) Current distribution in an over-voltage tolerant pad of a 0.18 μm technology obtained by circuit simulation. The pins (1) are connected to the I/O cell logic, and pin (2) is connected to the floating-well control circuitry. A small portion of the ESD current flows across N2 and leads to an electrical fail. b) Failure signature of the small NFET N2, which is not ESD optimized.

with ESD extensions. The main problem is the snap-back behaviour of the FETs, which causes numerical instabilities. Therefore, a good strategy is to substitute only those devices with ESD models which carry high currents or are exposed to high voltages. Practical limits could be set, such as the trigger voltage and the trigger current of the FETs. For the first run, a little of experience on ESD is helpful for choosing those elements in the circuit which must be considered as ESD relevant. Obviously, all elements that are connected directly to the pad or via a directed low-ohmic path ($< 500\ \Omega$ can be used as a rule of thumb) must be modelled by the corresponding ESD model. Further candidates are power clamps, which are in any case designed to shunt the current and limit the voltage during an ESD event. For this reason, the intended operation regime of power clamps is often far beyond the standard operating regime of transistors. This approach has been used to investigate an over-voltage tolerant I/O cell with about 70 active elements.

By ESD circuit simulation it is possible to analyse shunt paths that are not obvious at first sight. The current paths for a 2 kV stress are summarized in Figure 6.24. It is quite clear from this that the positive ESD stress to VDDP causes damage to the small NFET N2 with $W = 2$ μm, although the portion of the current flowing through this device is less than 4 % of the total ESD current. As intended by the design, most of the ESD stress is discharged through the protection element to VDDP. At first glance, the path "pad → p driver → NFET N2" seems to be

well protected by the shunt element. However, as soon as the voltage between pad
and VDDP exceeds approximately the trigger voltage of the N2 plus one diode
on-voltage (i.e. the on-voltage of the diode between $p+$ diffusion and n well of
p driver, approx. 0.7 V), N2 can trigger. The risk for N2 mainly depends on
the clamping capability of the protection element ESDVSS and on the discharge
current.

This case study clearly proves the benefits and capabilities of compact simulation. Even in the pre-silicon phase, ESD circuit simulation can be used as a guide
for ESD-robust designs. As the complexity of the pad circuits increases, there is
a high probability that even experienced ESD engineers may overlook "hidden"
shunt paths, which may result in low-threshold product failures.

Even using circuit simulation with sophisticated models, the detection of possible over-voltages and over-currents during the ESD event in order to find ESD-
weak paths/nodes in the design is a crucial problem. Detecting an over-voltage
is generally important for the breakdown of parasitic elements such as GOX, diffusions or inverters. An over-voltage can be flagged relatively easily by simply
comparing the voltage during the ESD event with the minimum breakdown voltage for the parasitic elements. In contrast, the detection of an over-current is much
more difficult. The problem is that the maximum withstand current of a device
depends mainly on its layout, information which is generally not available in the
netlist. For non-ESD-protected devices, as used in the core, the ESD robustness
is determined by the finger width of the device. In the above example, the current
through N2 is fatal for a device without ESD optimization ($I_{t2} \approx 1$ mA/µm), but
also for ESD-optimized NFETs ($I_{t2} \approx 15$ mA/µm). Assuming that the width of
N2 is doubled, an ESD optimized device would pass 2 kV, but a standard device
would certainly fail. A thorough analysis by an ESD engineer seems indispensable.
In order to allow fully automatic "flagging" of ESD endangered devices, a tool is
required which can assign to each device in the net list an ESD withstand current
by extracting the ESD-relevant information from the layout.

6.6   Limitations of ESD circuit simulation

For several years, ESD circuit simulation has been proving itself to be a powerful
tool for the design of ESD robust circuits. However, it should be remembered that
even with the current state-of-the-art compact models the field of application for
circuit simulation has some limitations.

On the one hand, circuit simulation always uses compact models with a strongly
simplified description of the device physics. The real behaviour of the devices under
ESD stress is in many cases determined by thermal effects, such as local self-heating
of the device, and 3D effects such as inhomogeneous turn-on or filamentation. Such
non-uniform effects can in general not be mapped accurately using a compact
model. If these effects are crucial for the behaviour of a circuit, mixed-mode
simulation using a 2D or 3D device simulator for the simulation of the critical
devices and a circuit simulator for the less critical elements could be one alternative
approach.

On the other hand, even if it is possible and physically reasonable to extend the compact models towards more realistic models, these approaches often degrade the numerical robustness. Unacceptable run times and, worse, numerical instabilities often result when models are physically "improved". In the authors' experience, numerical problems are often much more of a limiting issue than the deviation of simulation compared to the experimental data owing to the use of simplified models. Numerically stable models will be one of the challenges to be solved on the way towards a "full-chip simulation".

Another key issue that limits the applicability of today's circuit simulation is the complexity of the circuit which must be simulated and the unknown constraints for the simulations, such as the potential at nodes not directly in the stress path. As shown in Section 6.5, the potential at these nodes significantly influences the result of the simulation. Since floating nodes must be avoided for fear of numerical instabilities, and an analogue full-chip simulation is not possible with the current models and simulators, appropriate constraints must be chosen by the ESD engineer based on the often long-standing experience with critical circuits. This limits, in particular, the applicability of ESD circuit simulation for non-ESD experts.

Summary

The most important results of this chapter are summarized below.

- In ESD circuit simulation, the compact models must carefully be balanced between physical contents and numerical complexity. In general, the size of the circuits which can be simulated is inversely proportional to the complexity of the compact model.

- Compact models of ESD protection devices and active elements which are appropriate for ESD circuit simulation can be derived by means of rather simple "textbook physics".

  - The compact model for a diode can be based on an extension of Schottky's ideal diode equation taking into account non-ideal emission factors and serial resistances.

  - For the diffusion/well resistor, the ideal Ohm's law must be modified by introducing an effective mobility which considers velocity saturation effects.

  - A BJT can successfully be simulated by a simple Ebers–Moll model including additional avalanche sources. For the correct transient description diffusion/junction capacitances must be added.

  - For FETs, a modular compact model consisting of a BJT and a standard MOSFET model in parallel has several practical advantages, particularly the ability to use well-characterized standard MOSFET models. Depending on the application, further avalanche sources and diodes must be included in order to account for the different stress directions between the terminals.

The models must be extended to take account of special requirements, such as electro-thermal coupling for temperature-critical applications where self-heating effects play a dominant role, for example devices in SOI technologies.

- A crucial issue for a successful ESD circuit simulation is the parameter extraction procedure and the verification of the models. This can be either done by ESD device simulation studies or by the experimental analysis of appropriate test structures.

- A maximum current density through a device and a maximum voltage occurring at a node of the circuit can serve as appropriate failure criteria for the ESD circuit simulation.

- A typical application of ESD circuit simulation treats circuits up to several tens devices like I/O cells or small analogue macros connected to a pad. For these cases it has been shown that discharge paths can be detected and the circuit can be optimized by means of ESD circuit simulation.

Bibliography

Amerasekera A., Chang M. C., Seitchik J., Chatterjee A., Majaram K., Chen J. C., "Self-heating Effects in Basic Semiconductor Structures", IEEE Electronic Devices (1993), 1836.

Amerasekera A., Duvvury C., *ESD in Silicon Integrated Circuits, Second Edition,* John Wiley, Chichester, England, 2002.

Amerasekera A., Ramaswamy S., Chang M. C., Duvvury C., "Modeling MOS Snapback and Parasitic Bipolar Action for Circuit-Level ESD and High Current Simulations", Proc. IRPS (1996), 318.

Beebe S., "Simulation of Complete CMOS I/O Circuit Response to CDM Stress", Proc. 20th EOS/ESD Symposium (1998), 259.

Blicher A., *Field-Effect and Bipolar Power Transistor Physic,* Academic Press, New York, 1981.

Boselli G., Ramaswamy S., Amerasekera A., Mouthaan T., Kuper F., "Modelling Substrate Diodes under Ultra-High ESD Injection Conditions", Proc. 23th EOS/ESD Symposium (2001), 71.

Carbajal B., Cline R., Andersen B., "A Successful HBM ESD Protection Circuit for Micron and Sub-Micron Level CMOS", Proc. 14th EOS/ESD Symposium (1992), 234.

Chynoweth A. "Uniform Silicon p-n Junctions II. Ionization Rates of Electrons", J. Appl. Phys. **31** (1960), 1161.

Diaz C., Kang S., Duvvury C., "Circuit-level Electrothermal Simulation of Electrical Overstress Failures in Adanced MOS I/O Protection Devices", IEEE Trans. CAD **13**(1994), 482.

Dutton R. W. "Bipolar Transistor Modeling of Avalanche Generation for Computer Circuit Simulation", IEEE Trans. Electr. Devices **22** (1975), 334.

Ebers J. J., Moll J. N., "Large-Signal Behavior of Junction Transistors", Proc. IRE **42** (1954), 1761.

Gao X. F., Liou J. J., Ortiz-Conde A., Bernier J., Croft G., "A Physics-based Model for the Substrate Resistance of MOSFET's", Solid State Electr. **46** (2002a), 853.

Gao X. F., Liou J. J., Bernier J., Croft G., Ortiz-Conde A., "Implementation of a Comprehensive and Robust MOSFET Model in Cadence SPICE for ESD Applications", IEEE Trans. CAD **21** (2002b), 1497.

Gieser H., Haunschild M., "Very-Fast Transmission Line Pulsing of Integrated Structures and the Charged Device Model", Proc. EOS/ESD Symposium (1996), 85.

Gummel H. K., Poon H. C., "An Integral Charge Control Model of Bipolar Transistors", Bell Syst. Tech. Journal **49** (1970), 827.

Hsu F.-C., Ko P.-K., Tam S., Hu C., Muller R. S., "An Analytical Breakdown Model for Short-Channel MOSFET's", Proc. EOS/ESD Symposium (1996), 85.

ISE Integrated Systems Engineering AG, Switzerland, ISE TCAD Manuals, 1998, Release 5.0, Part 14: DESSIS.

Joshi S., Rosenbaum E., "Compact Modeling of Vertical ESD Protection npn Transistors for RF Circuits", Proc. 24th EOS/ESD Symposium (2002), 289.

Juliano P., Rosenbaum E., "A Novel SCR Macromodel for ESD Circuit Simulation", Int. Electron Devices Meeting Tech. Dig.,(2001), 319.

Kittel C., *Introduction to Solid State Physics, 6th Edition,* John Wiley & Sons, New York, 1983.

Kurimoto K., Yamashita K., Miyanaga I., Hori A., Odanaka S. "An Electrothermal Circuit Simulation Using an Equivalent Thermal Network for Electrostatic Discharge (ESD)", Proc. VLSI Techn. Symposium (1994), 127.

Li T., Suh D., Ramaswamy S., Bendix P., Rosenbaum E., Kapoor A., Kang S. M., "Study of a CMOS I/O Protection Circuit Using Circuit-Level Simulation", Proc. IRPS (1997), 333.

Li T., Tsai C.-H., Rosenbaum E., Kang S.-M., "Substrate Resistance Modeling and Circuit-level Simulation of Parasitic Device Coupling Effects for CMOS I/O Circuits under ESD Stress," Proc. 20th EOS/ESD Symp. (1998), 281.

Lim S. L., Zhang X., Beebe S., Dutton R. W. "A Computationally Stable Quasi-Empirical Compact Model for the Simulation of MOS breakdown in ESD protection Circuit Design", Proc. SISPAD (1997), 161.

Luchies J. R. M., de Koert C. G. C. M., Verweij J. F. "Fast Turn-on of an NMOS ESD Protection Transistor; Measurements and Simulations", Proc. 16st EOS/ESD Symposium (1994), 266.

Ma C. L., Lauritzen P. O., Sigg J., "Modeling of Power Diodes with the Lumped-Charge Modeling Technique", IEEE Transactions on PowerElectronics, **12** (1997), 398.

Massobrio G., Antognetti P., *Semiconductor Device Modeling with SPICE,* McGraw-Hill, New York, 1993.

Mergens M., Wilkening W., Mettler S., Wolf H., Stricker A., Fichtner W., "Analysis and Compact Modeling of Lateral DMOS Power Devices Under ESD Stress Conditions", Proc. 21st EOS/ESD Symposium (1999), 1.

Mergens M., Wilkening W., Mettler S., Wolf H., Fichtner W., "Modular Approach of a High Current MOS Compact Model for Circuit-level ESD Simulation Including Transient Gate-coupling Behavior", Proc. IRPS (1999a), 167.

Mergens M., Wilkening W., Kiesewetter, G., Mettler S., Wolf H., Hieber J., Fichtner W., "ESD-level Circuit Simulation – Impact of Gate RC-Delay on HBM and CDM robustness", Proc. 22nd EOS/ESD Symposium (2000), 446.

Mergens M., "On-Chip ESD Protection in Integrated Circuits: Device Physics, Modeling, Circuit Simulation", PhD thesis, ETH Zurich, 2001; published Hartung–Gorre-Verlag, Konstanz, 2001.

Miller S. "Ionization Rates for Holes and Electrons in Silicon", Phys. Rev. Lett. **105** (1957), 1246.

Muller R. S., Kamins T. I., *Device Electronics for Integrated Circuits, 2nd Edition,* John Wiley & Sons, New York, 1986.

Puvvada V., Duvvury C., "A Simulation Study of HBM Failure in an Internal Clock Buffer and the Design Issue for Efficient Power Pin Protection Strategy", Proc. 20th EOS/ESD Symposium (1998), 104.

Puvvada V., Srinivasan V., Gupta V., "A Scalable Analytical Model for the ESD N-Well resistor", Proc. 22th EOS/ESD Symposium (2000), 437.

Raha P., Ramaswamy S., Rosenbaum E., "Heat Flow Analysis for EOS/ESD Protection Device Design in SOI Technology", IEEE Trans. Electr. Dev. (1997), 464.

Ramaswamy S., Li E., Rosenbaum E., Kang S.-M. "Circuit-level Simulation of CDM-ESD and EOS in Submicron MOS Devices", Proc. 18th EOS/ESD Symposium (1996), 316.

Russ C., Gieser H., Egger P., Irl S. "Compact Electro-Thermal Simulation of ESD-Protection Elements", Proc. ESREF (1993), 395.

Russ C., Verhaege K., Bock K., Roussel P. J., Groesenecken G., Maes H. E. "A Compact Model for the Grounded-Gate nMOS Behaviour under CDM ESD Stress ", Proc. 18th EOS/ESD Symposium (1996), 302.

Russ C., *ESD Protection Devices for CMOS Technologies: Processing Impact,Modeling, and Testing Issues,* PhD thesis, Technical University Munich, 1999; published Shaker-Verlag, Aachen, Germany, 1999.

Shockley W., *Electrons and Holes in Semiconductors,* Van Nostrand, Princeton (N.J.), 1950.

Stricker A.D., Mettler S., Wolf H., Mergens M., Wilkening W., Gieser H., Fichtner W. "Characterization and Optimization of a Bipolar ESD-Device by Measurements and Simulations", Proc. 20nd EOS/ESD Symposium (1998), 290.

Sze S. M., *Physics of Semiconductor Devices, 2nd Edition,* John Wiley & Sons, New York, 1981.

Sze S. M., *Modern Semiconductor Device Physics,* John Wiley & Sons, New York, 1997.

Verhaege K., Russ C., Robinson-Hahn D., Farris M., Scanlon J., Lin D., Veltri J., Groesenecken G. "Recommendations to Further Improvements of HBM Component Level Test Specifications", Proc. 18th EOS/ESD Symposium (1996), 40.

Wang Y., Juliano P., Joshi S., Rosenbaum E., "Electrothermal Modeling of ESD Diodes in Bulk-Si and SOI Technologies", Proc. 22nd EOS/ESD Symposium (2000), 430.

Wolf H., Gieser H., Wilkening W., "Pulsed Thermal Characterisation of a Reverse Biased PN-Junction for ESD HBM Simulation", Microel. Rel. **36** (1996), 1711.

Wolf H., Gieser H., Stadler W., "Bipolar Model Extension for MOS Transistors Considering Gate Coupling Effects in the HBM ESD Domain ",Proc. 20nd EOS/ESD Symposium (1998), 271.

Wolf H., Gieser H., Wilkening W., "Analyzing the Switching Behavior of ESD-Protection Transistors by Very-Fast Transmission Line Pulsing,", J. Electrostatics **49** (2000), 111.

Wolf H., Gieser H., Stadler W., Esmark K., "ESD Circuit Simulation for the Prevention of ESD Failures - Application to Products in a 0.18 m Technology", Proc. IRPS (2001), 430.

Chapter 7

Chip-level ESD verification

The difficulty of ESD protection development stems not only from the complexity of the underlying physics, but also from the complexity of the "value chain", stretching from the choice of optimum technology parameters through to the correct implementation of the ESD measures in the final product. In this chapter the challenge of the last stages of the race, namely checking a complete IC with, eventually, several tens of millions of gates and many hundreds of I/O pins, will be addressed. As explained in the introduction, ESD protection relies on the power rails as the "lightning conductor" around the core, where the discharge current can be conducted to the grounded pin without generating a high voltage drop. The tendency of modern ICs to break the power network into many power domains, for functional reasons, is counterproductive in this case. The protection path has to be evaluated very carefully. Even then, constellations might occur where even elements in the core need special ESD treatment. Since all the effort put into the verification and optimization of the ESD protection concept finally fails if these issues are overlooked, the ESD development engineer must spend a significant part of his/her time checking the schematics and layouts of complete IC designs. Naturally, this "manual" method of ESD design review requires a significant effort for each design project, and is prone to mistakes. An automated verification is the essential goal, both for more reliable design and for more efficiency. For this purpose there are two basic approaches which will be discussed in this chapter: formal verification and electrical simulation at chip level.

7.1 Goals

Let us first review what the tasks of a chip-level ESD verification should be. The predominant issues in an ESD review of a complete IC are the coupling of the power domains and the placement of supply cells, also including package issues like the use of double-bonded pins and the availability of power rings in the package. There must be at least one well-defined discharge path between two arbitrary power domains, which means that a supply line must either be fed through metallically, or coupled by antiparallel diodes, or connected via a double-bond on a common

243

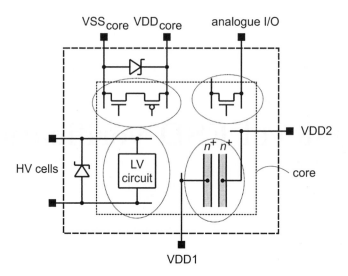

Figure 7.1: Particular critical locations in a chip core which must be analysed in order to
avoid ESD fails.

pin of the package (see Section 2.5). All this information (including the bonding of
the power cells) must be provided to the tool and be checked for correctness. This
requires the calculation of the voltage drop in the power network, including the
characteristic of the power clamp and the coupling diodes, and the extraction of
the resistance of the metal bus. A warning should be given, if the voltage exceeds
a given critical value. Actually, to allow an optimized floor planning of the IC it
is favourable to perform this analysis in an early design phase before the complete
chip layout is almost finished and it becomes difficult to implemented any changes
in the supply concept.

In certain constellations like critical inverter stages and parasitic npn structures
between different supply domains (Duvvury, 1988; Johnson, 1993; Puvvada, 1998)
it is necessary to detect possible failure locations in the core, even below an overall
critical voltage limit. This type of failure mechanism must be probed in addition
to the voltage evaluation of the power network (Figure 7.1).

Another common problem is the erroneous mismatch between a protection
concept implemented in an I/O cell and the protected circuitry, such as a PLL with
low voltage transistors connected directly to a pad using a high voltage protection.
As during place and route only high level views of the I/O cells are used, the design
engineer may not recognize, except by checking the written documentation, which
devices and analogue circuits fit with the chosen interface cell. An automated
verification should find all devices connected to the pad via critical paths and
verify whether the protection concept is compliant.

A major issue for analogue designs is the implementation of ESD design rules
and ESD circuit measures in "non-I/O" circuit blocks. For example, diffusions
belonging to a transistor of the "core" area, which is connected to the pad via
low ohmic paths, must commonly follow ESD specific design rules. This can be

checked at the level of each macro, but should also be verified at the chip level as a final check.

To achieve these goals, the verification can be based on a formal verification procedure for checking the layout and schematics, an electrical simulation of the ESD event, or a combination of both.

7.2 Verification methods

The formal verification by performing basic ESD design rule checks is a well known approach (Kang, 1998; Bass, 2000). However, in its simplest form it can only check devices marked as ESD-relevant during the design phase, for example by ESD marking layers, concerning straightforward design rules like spacings. The more critical case where the designer has overlooked ESD-relevant devices is not covered. A more sophisticated approach is to include information about the connectivity of the devices and to automatically select all ESD-relevant devices by an extraction step (Sinha, 1998). By also including the evaluation of the bus resistance, many possible failures can be detected. Numerical methods have been applied for extraction of the voltage drop in the power supply protection network more precisely than in the layout based verification (Ngan, 2001).

However, there is a limitation with formal verification, which can only work with pre-defined failure mechanism in the form "consider all diffusions connected to a bond pad via a resistance below x Ω as ESD sensitive". It cannot discover new critical discharge paths, such as those arising from transient powering of the IC during the ESD pulse. To fully incorporate the effects of the ESD discharge, an ESD compact simulation, as discussed in the previous chapter, is necessary. This would cover, in conjunction with a design rule check (DRC), all the needs listed above. Particularly for CDM, a reasonable ESD evaluation of the IC inevitably requires a chip-level ESD simulation, owing to the dependence of the waveform and the peak current on the on-chip and package capacitance. In this case the package parameters like capacitance to a ground plane must also be included. The main stumbling block for such a project is the complexity of the required simulation. Even under normal operational conditions a full chip compact simulation is not applicable to designs with 100 million gates or more. Beyond this, the ESD-relevant models for the compact simulation are even more complex, including the snapback behaviour of the bipolar transistors and thermal effects. Finally, a comprehensive ESD simulation needs the calculation of many different pin combinations requiring hundreds to thousands of simulation cycles.

The challenge is now to simplify the simulation without eliminating the critical paths and effects. This task has been identified, but has not yet solved; only a few contributions on this topic have been published so far (Baird, 2000; Lee, 2000; Venugopal, 2002; Streibl, 2003).

7.3   First steps to chip-level ESD simulation

Following the approach of Streibl (2003), an example is discussed how an ESD fail can be extracted out of the multiple combination of possible discharge paths by a chip-level simulation. One part of the task is to find the circuit representing the essential features of the IC under ESD, the other task is to model the various discharge combinations.

7.3.1   Extraction of an ESD netlist

As discussed in Section 1.3 there are in principle two different failure mechanism which have to be considered when an IC is ESD stressed. One is the local damage of devices which are directly connected to signal pads. This typically involves only a few devices e.g. the driver stage of an I/O cell and can usually be treated by the compact simulation discussed in Chapter 6. The second involves the discharge path via the supply lines and leads to fails of devices in the core region or at positions possibly far from the node where the ESD pulse is forced into the circuit. This behaviour requires the consideration of the complete IC and is the topic for the chip-level simulation. As the supply lines feed the stress pulse into the IC, it is obvious to consider first the supply protection concept including the power clamps, the coupling elements between the supply domains, connections via the package, e.g. double bonds, and the bus resistance. Under the assumption that a discharge path across core elements (or parasitics) does not significantly modify the current distribution in the supply net below irreversible damage, simple macro ESD models for critical "core" elements can then be attached to the netlist of the supply network. Here, both elements at the interface between power domains including possible substrate parasitics, e.g. between neighbouring n guardrings connected to different supplies, and critical structures between the power rails of a single power domain, say, critical inverter configurations, have to be included.

It has been proposed in literature to use a preselection procedure of critical pathes to limit the number of nodes in the final ESD simulation (Baird 2000; Streibl 2003).

The assumption of negligible influence of the core on the current distribution and voltage profile of the supply network, is not valid for CDM discharges where the capacitance of the single IC domains strongly influences the waveform (Lee, 2000). In principle this is also an issue for HBM discharge, if the rise time determines the triggering of internal pathes, and a simulation of the transient current distribution and voltage profile in the supply network including the RC net connected to the rails is required. This significantly complicates the simulation due to the large increase of nodes and the extraction of an appropriate RC net. However, for the extraction of the worst case conditions for HBM (and MM) a quasi-dc evaluation is sufficient, where the capacitance can be neglected (Streibl, 2003).

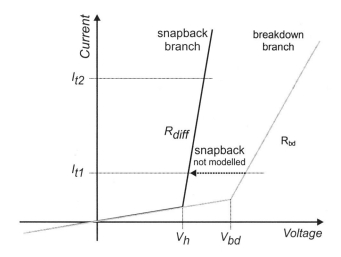

Figure 7.2: ESD and circuit devices are modelled either in the breakdown or snapback branch of their IV characteristic. The snapback mechanism itself is not described. Instead simulation runs for both possible IV branches are performed in a permutational sequence. The parameters are explained in Table 7.1.

7.3.2 ESD Modelling

Typically, the number of nodes which have to be considered for an IC with several power domains is in the range of several hundred to several thousands even for the simplified ESD netlist discussed above. As a large number of stress combinations between the various pins have to be rehearsed (typically several thousands for HBM or MM), it is important that a fast and stable numerical calculation of each simulation run is ensured. This mainly depends on the used ESD models. While the snapback of ESD protection and circuit devices is one of the key issues to take care of in ESD simulation, it makes the numerical treatment extremely unstable and computing intensive. Beside this, highly complex and delicately calibrated models must be extracted for a correct modelling. Fortunately, for the analysis of a worst case situation during the ESD discharge a very much simplified IV characteristics can be used neglecting the snapback. Streibl (2003) describes all elements which snapback has to be considered by a bimodal IV characteristic, where each branch is monotonic (Figure 7.2). One of the states reproduces the breakdown characteristic of the device without triggering the bipolar transistor, i.e. without snapback, the other describes the high current branch of the device after the snapback. In the simplest version both characteristics consist of a high-ohmic regime modelled by a linear IV dependence which changes to a low-ohmic behaviour beyond the breakdown voltage resp. the holding voltage. For a given constellation of states of the (bimodal) devices the current and voltage in the network can be calculated by standard circuit simulators.

In the simulation process both states of the bimodal characteristic are examined and the worst case constellation, which is in accordance with the given physical

Table 7.1: Description of ESD parameters of Figure 7.2

Parameter	Explanation
V_{bd}	breakdown voltage
R_{bd}	differential resistance in the breakdown branch
I_{t1}	trigger current of bipolar element
V_h	holding voltage
R_{diff}	differential resistance in the low ohmic state of the device
I_{t2}	current level that cause destruction of the protection element under square pulse conditions

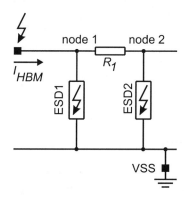

Figure 7.3: Simple ESD circuit for clarification of the permutational ESD simulation procedure.

constraints, is extracted. To treat all the permutations which can occur in the complex circuit described by the ESD netlist numerical methods like complete enumeration (Streibl, 2003) or shortest path algorithms (Ngan, 2001) have to be applied. This approach also ensures that the problem of competing pathes discussed in chapter 6 is automatically covered. A damage is detected, if the critical current I_{t2} or the upper limit voltage of the ESD design window is exceeded.

This procedure is illustrated by a simple example of a digital input stage (Figure 7.3)

ESD1 and ESD2 are identical ESD protection devices (e.g. ggNFETs) with the ESD parameters given in Table 7.2. The resistance R_1 is chosen to 5 Ω. An ESD current of $I_{HBM} = 1.3$ A is applied to the circuit at node 1 versus VSS. To describe both ESD element, ESD1 and ESD2, in their two breakdown parameter sets (V_{bd}, R_{bd}) and (V_h, R_{diff}), $2^2 = 4$ simulation runs are required.

Table 7.3 lists the simulation results for each possible permutation, with the number in the "permutation index" column indicating in binary form the chosen IV parameter set of ESD1 (first digit) and ESD2 (second digit). A "0" in the permutation index stands for the breakdown parameter (V_{bd}, R_{bd}) set and a "1"

Table 7.2: ESD parameters for the example in Figure 7.3.

Parameter	Value	Parameter	Value
V_{bd}	10 V	V_h	5 V
R_{bd}	20 Ω	R_{diff}	2 Ω
I_{t1}	10 mA	I_{t2}	1.3 A

Table 7.3: Simulation results for the example in Figure 7.3.

permutation index	node 1 voltage	ESD1 current	node 2 voltage	ESD2 current
~~(00)~~	~~24 V~~	~~0.73 A~~	~~21 V~~	~~0.57 A~~
~~(01)~~	~~13 V~~	~~0.15 A~~	~~7.3 V~~	~~1.15 A~~
(10)	7.6 V	1.3 A	7.6 V	0
(11)	7.0 V	1.0 A	5.6 V	0.3 A

for the snapback parameter set (V_h, R_{diff}).

As expected, the node voltages differ significantly between the permutations. Before discussing the simulation result unrealistic and non physical simulation runs have to be sorted out: Permutation (01) with ESD1 in breakdown and ESD2 in snapback mode leads to a current of 0.15 A through ESD1, which is beyond its trigger current point, so ESD1 should rather be modelled in the snapback mode. This case, however, is already included by the simulation run of permutation (11). Therefore simulation run (01) can be marked as "invalid" run and be dropped. Same applies to run (00), leaving the two valid and physically sensible simulation runs (10) and (11). In real world both results might occur, and the outcome would be determined by minor technological differences between the two ESD devices or by transient effects influenced by layout and circuit constellations. As for ESD analysis one is interested in the worst case situation, a maximum voltage drop of 7.6 V at node2 for the HBM event is derived from this permutational simulation procedure.

While the simulation result for the simple demonstrator circuit itself is not astonishing and can also be deduced without employing a computer, this simulation approach is applicable in the same way to circuits of much higher complexity. However, as the number of necessary simulation runs for N snapback elements increases with 2^N it might become necessary to limit their number. Therefore, for a complex circuit of an IC a simulation a pre-run procedure is proposed preceding the permutation analysis, to determine among all circuit and snapback devices the ones that might be triggered into snapback for a given stress pin combination (Streibl, 2003). With this method e.g. a circuit containing 600 elements and 300 nodes has been simulated for 1600 different stress combinations on a 360 MHz RISC processor within 8 hours (Streibl, 2003). In this example, an error in the power supply concept due to wrong bonding has been detected and the failure position has been localized.

Of course, this method is also applicable to complex analog circuits which can not be treated by compact simulation due to numerical problems.

While the simulation of the mere power supply concept is accessible, the simulation of a ESD netlist representing the complete IC still lacks of the automated extraction of the critical core elements.

7.4   Future Challenges for a chip-level ESD simulation

As no satisfying method of the comprehensive chip-level ESD simulation is yet available, the focus of the following discussion will be on the required features to provide a guide for the further development.

The input data for a chip-level simulation can be based either on the schematics, the layout, or an ESD-specific circuit representation (ESD views). As, for example, the resistance in the metal bus plays an important role, information must be provided about the extracted resistances and, particularly for CDM events, the capacitances. This requires as input either the physical layout or a back-annotated schematic. To include all possible failure locations, parasitic elements in the substrate have also to be extracted. This makes information about the physical layout an inevitable input for a full-blown chip-level analysis.

The simulation itself requires consideration of the IV characteristics of the protection circuitry and an evaluation of critical over-voltage and current overstress. This should be based on the available circuit simulators. Any devices exhibiting a snapback especially if they are part of the protection circuit have to be modelled accordingly. The models of the protected circuit, like the core logic, do not in general have to be so elaborated. The overwhelming portion of the circuit just acts as capacitance between the power rails during the ESD event. However, certain parts are ESD critical and have to be considered more carefully. Any device at the interface between different power domains has to be modelled at least as diode with a critical breakdown voltage. Inverter-like structures, where gate control of a large PFET during ESD leads to sufficient current for damaging a serial NFET, are even more difficult to include consistently. In addition to the critical paths found in the schematics, parasitic paths like substrate npn transistors between neighbouring circuit blocks must also be added in order to evaluate all possible ESD damages. To be able to perform a simulation considering all these constraints, the generation of an "ESD netlist" will be inevitable. So that the parasitic elements can be included, an extraction procedure has to precede the simulation.

The simulation should cover the pin combinations appearing in a real stress test according to the test standards (see e.g. JEDEC–HBM 2000. As such, it should act as "virtual" ESD test. Typically for higher pin counts (> 8) not all possible pin combinations are stressed in a hardware test, but the pins are assorted into stress groups to reduce the test effort. An equivalent approach should be chosen for the simulation. This is, all pins are stressed to the different supplies and, in addition, each single I/O pin is discharged to the rest of the I/O pins grounded during an HBM or MM investigation. For a CDM investigation the amount of charge and its distribution on the packaged IC must be modelled first. The discharge itself is done by hard-grounding each pin subsequently or – more realistically – simulating the air discharge between each tested pin of the IC and the grounded pin of the tester.

Finally, after having built up such a complex simulation routine, it is surely preferable to migrate it easily from technology to technology. This can be done by putting all the technology-specific information in an ESD TechFile which is used during the extraction of the parasitics and the generation of the ESD netlist, keeping the procedure itself unchanged.

The possible flow of an HBM chip-level verification is summarized in Figure 7.4. Here the hierarchical approach of the simulation is important. Before the chip-level simulation is performed, all ESD-critical single macros, such as I/O cells, should separately be evaluated by an ESD design rule check and eventually a circuit simulation including the ESD compact models. All internal weaknesses in the cells need to be sorted out on this level. The ESD design rules for elements marked as ESD critical must be correct. To proceed with the chip-level simulation, the protection network of the supplies including its resistance tree must be determined. ESD critical core elements and circuits have to be detected and the critical parasitic elements at the border of the macros should be extracted. The extraction procedure makes use of the ESD TechFile. An ESD netlist containing the power protection network including the bus resistance, the critical core elements and the parasitic elements is generated. A test set-up for the stress simulation is defined which covers all the relevant pin combinations and stress levels. Then the chip-level simulation can be performed for this netlist using the ESD macro models of I/O cells, supply protection elements, core circuitry and parasitics. Based on the pre-selected electrical failure criteria of excess voltage and/or current, the error output provides the location of possible fails together with the critical discharge path. After possible correction of the design a new iteration of ESD design rule check and ESD simulation has to be performed.

Extraction procedure, netlist generation, simulation and error output should all be part of the standard design flow, so as to guarantee consistency of the data and to facilitate the application for the design engineer. The handling of the large amount of data for the chip-level simulation requires efficient data management, and the necessary access to the layout data and schematics would also suggest integration with the standard design system.

The currently applied chip-level verification methods, as discussed above, already sort out a large number of the typical problems. However, any remaining errors will lead to a redesign causing costs for reticle sets in the 1 Mio Dollar regime and a significant delay in the shipping of the product. Therefore, the goal of "first time right" requires – and justifies – the effort to solve the problems of a comprehensive ESD chip-level verification.

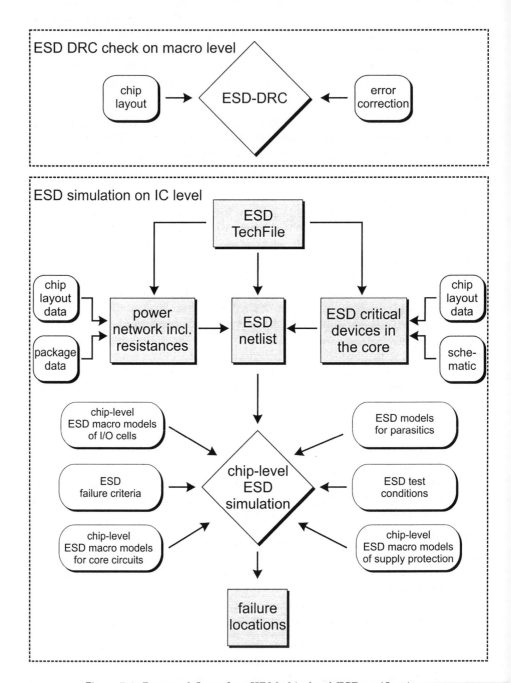

Figure 7.4: Proposed flow of an HBM chip-level ESD verification.

Summary

- A final evaluation of the ESD protection measures on the complete IC is required in order to avoid errors in the integration of all the different ESD protection components. This mostly affects the arrangement in the power supply concept, and may involve problems such as missing coupling between power domains, or unwanted parasitic elements due to placement of the different circuit parts.

- As this final check also includes core circuitry and numerous I/O and supply cells, a straightforward circuit simulation is not applicable because of the likely numerical problems.

- The problem of extracting more simplified (ESD) models which can be treated within a chip-level ESD simulation without neglecting the ESD critical issues has not yet been solved. In this chapter, the requirements for such a task have been collected, and a proposal for the principle of a chip-level ESD simulation flow has been made.

Bibliography

Baird M., Ida R., "VerifyESD: A Tool for Efficient Circuit Level ESD Simulation of Mixed-Signal ICs", Proc. EOS/ESD Symposium (2000), 465.

Bass S., Nickel D., Sullivan D., Voldman S., "Method of Automated ESD Protection Level Verification", US Patent 6086627 (2000).

Duvvury C., Roundtree R., Adams O., "Internal Chip ESD Phenomena Beyond the Protection Circuit ",IEEE Trans. Electr. Dev. (1988), 2133.

Johnson C., Qawami S., Maloney T., "Two Unusual HBM ESD Failure Mechanisms on a Mature CMOS Process",Proc. EOS/ESD Symposium (1993), 225.

JEDEC Solid State Technology Association, *Electrostatic Discharge (ESD) Sensitivity Testing Human Body Model (HBM)*, JESD22–A114–B, 2000.

Kang S-M.S., Duvvury C., Diaz C.H., Ramaswamy S., "Methods, Apparatus and Computer Program Products for Synthesizing Integrated Circuits with Electrostatic Discharge Capability and Connecting Ground Rules Faults therein", US Patent 5796638 (1998).

Lee J., Huh Y., Chen J.W., Bendix P., Kang S.-M., "Chip-Level Simulation for CDM Failures in Multi-Power ICs", Proc. EOS/ESD Symposium (2000), 456.

Ngan P., Gramacy R., Wong C.-K., Oliver D., Smedes T., "Automated Layout Based Verification of Electrostatic Discharge Paths", Proc. EOS/ESD Symposium (2001), 96.

Puvvada V., Duvvury C., "A Simulation Study of HBM Failure in an Internal Cluck Buffer and the Design Issue for Efficient Power Pin Protection Strategy", Proc. 20th EOS/ESD Symposium (1998), 104.

Sinha S., Swaminathan H., Kadamati G., Duvvury C., "An Automated Tool for Detecting ESD Design Errors",Proc. EOS/ESD Symposium (1998), 208.

Streibl M., Zängl F., Esmark K., Schwencker R., Stadler W., Gossner H., Drüen S., Schmitt-Landsiedel D., "High Abstraction Level Permutational ESD Concept Analysis" accepted for publication EOS/ESD Symposium 2003.

Venugopal P., Sinha S., Ramaswamy S., Duvvury C., Prasad G., Raghu, Kadamati G., "Integrated Circuit Design Error Detector for Electrostatic Discharge and Latch-up Applications", US Patent 6493850B2 (2002).

Chapter 8

Outlook

In this book, we have shown how, from starting with basic information about the process, the ESD robustness of an IC can successfully be evaluated by the application of the presented simulation flow. A large number of critical issues, ranging from the second breakdown of the protection devices to the transient behaviour of complete I/O cell circuits, can be dealt with in this way. However, there are still open questions which need the attention of the research community.

Currently the major restrictions are seen in

- the high numerical effort required for 3D process and device simulation;

- the limited quality of the profiles and the topology in 3D process simulations;

- the unsatisfying situation concerning the thermal component in the circuit simulation;

- the missing automated extraction procedure of critical circuits and parasitics for a comprehensive IC level ESD simulation.

These constraints must be considered in the light of the challenges faced by ESD protection development in the future. The ESD design window becomes narrower and narrower. Already at the 90 nm node the gate oxide breakdown voltage under ESD is only 1 V above the gated breakdown of the drain junction. Following the scaling rules, the electrical thickness of the gate dielectric has to be reduced to an oxide thickness below 1 nm for a gate length of 20 nm. The resulting extremely low breakdown voltage of the dielectric makes it very difficult to protect such devices against ESD. Fortunately, it seems that other reasons, such as huge gate leakage and process control problems, will prevent this approach anyway. A solution is seen in the so-called high-k dielectrics, where the increase in electrical permittivity allows the application of thicker layers. However, it is not at all clear how they will behave under ESD conditions. As these dielectrics lead to a lower mobility – at least until now – they are thought of in conjunction with strained silicon, which would enhance the mobility. Here, silicon is grown on top of a SiGe buffer layer with a different lattice constant. The electrical properties of the

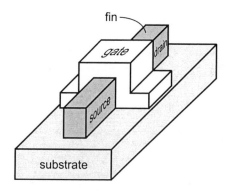

Figure 8.1: Schematic structure of a FIN-FET.

thin Si layer are modified as a result of the strain on the misadjusted lattice. The question arises of what happens under ESD conditions when the temperature in the strained lattice locally reaches very high values. To shrink the device dimensions even further will lead to completely new process steps. Fully depleted silicon-on-insulator (FD-SOI) technologies are on the point of being developed for mass production. Here the "bulk" silicon, which mainly carries the ESD current in most of the devices, has a thickness of only a few nanometers, see discussion in Chapter 5.5.5. A similar condition appears in FIN-FET (Figure 8.1), another option in the race to manufacture the smallest silicon FET devices, where the active silicon region is constructed as a vertical silicon fin covered by a double gate around the middle part. The two end pieces of the fin serve as source and drain region. Also here the volume of the "bulk" is very small. In both devices a greatly increased energy density is expected. Apart from this, some devices that worked well as ESD protection devices in the past, such as the STI bound diode, are no longer available in these technologies.

All these questions are typical topics for the future of process and device simulation. However, to address them, the physical models for the new materials and the small dimensions of a few 10 nm have to be evaluated. This is another challenge for basic research.

With the progress in technologies, not only the feature size will become smaller, but also the speed of the ICs will increase, also at the IC signal interfaces. Currently, hf signalling of 1–2.5 GHz is standard in mobile products; values up to 40 GHz are realized for optical network components. Both BiCMOS and CMOS technologies are employed in this field. Any ESD related measures like protection elements connected to the pad or serial resistance in the signal path are seen as degrading the hf performance. For mobile products, there is often a compromise for hf pins with extreme hf performance whereby the ESD requirement is lowered. To find the optimum between hf performance and ESD protection, circuit level simulations are the right tool. This is already possible with the existing tools; however, good hf models of the ESD protection elements also have to be extracted.

For more than a decade, CDM has been regarded as the most critical model in reflecting the field stress encountered by today's ICs. It is included in the qualification procedure of most of the semiconductor companies. However, knowledge of the behaviour of single devices and the circuits during these very fast pulses is still limited, mainly because of the absence of a characterization method in the past. Now the vf-TLP technique allows access to the parameters governing this time regime. Both device simulation models and circuit parameters will have to be refined based on the results of such investigations.

The starting point for this book was the conviction that simulation is a valuable – possibly in a few years even an indispensable – tool for optimization of the ESD protection of ICs. Therefore, as a final remark we would like to motivate the tool providers, both TCAD and EDA vendors, to support this topic. Only by continuous optimization of the tools, taking advantage of the steady progress made by the ESD research community, can a widespread, successful application be achieved. Keeping in mind that ESD verification is still the most critical topic before the qualification of a product, the savings, if a "first time right" design can be achieved, are enormous. Our vision is that compact simulation and chip-level ESD simulation/verification methodology will become an inherent part of the standard design flow. When that happens, a "virtual ESD test" will be possible within the standard sign-off procedure before tape-out. In this sense, press the button ... and good luck.

Appendix

Symbols

A	area (of depletion region, device cross section, ...)
C	concentration
C_{back}	background capacitance used in CDM models
C_{CDM}	capacitor used in the equivalent circuit of a CDM discharge
C_{chip}	chip capacitance
C_{dbe} (C_{dce})	diffusion capacitances of base–emitter (collector–emitter) junction
C_{ESD}	ESD capacitor
C_{HBM}	HBM capacitor (100 pF for component-level tests)
C_{jbe} (C_{jce})	junction capacitances of base-emitter (collector–emitter) junction
C_{jbe0} (C_{jce0})	zero-bias junction capacitances of base-emitter (collector–emitter) junction
C_{MM}	MM capacitor (200 pF for component-level tests)
C_{p}	specific heat
C_{s}	stray capacitance of ESD tester
C_{CDM}	capacitor used in the equivalent circuit of a SDM discharge
C_{tb}	test board capacitance of ESD testers
C_{th}	thermal capacitance
d	thickness of sample used in BLI experiment
D	diffusivity of impurities (intrinsic diffusion regime)
DCG	drain contact-to-gate spacing
D_{n} (D_{p})	electron (hole) diffusion constant
d_{Si}	thickness of silicon layer in SOI
D_{th}	thermal diffusion constant
E, $\vec{E}$	electric field
E_{g}	bandgap energy
E_{i}	threshold energy to generate an electron-hole pair
E_{op}	optical phonon energy
f_{T}	threshold frequency
G	generation rate

H	dissipated energy
I	general: current
I_{avc}	avalanche generation current between collector and base
I_{ave}	avalanche generation current between emitter and base
I_{avsb}	avalanche generation current between source and bulk
I_b	base current
I_c	collector current
I_{cc}	collector current through current source in EM model
I_{ce}	emitter current through current source in EM model
I_{ct}	current through current source in EM model ($= I_{cc} - I_{ce}$
I_d	drain current
I_{device}	current through the DUT during an ESD stress
I_e	emitter current
I_{ESD}	ESD current
I_{fw}	current at which a protection device is turned-on over the full width
I_h	holding current
I_{HBM}	HBM current during ESD test
I_{leak}	leakage current
I_{mf}	current which guarantees multi-finger triggering
$I_{sat,bc}$	saturation current of base-collector diode
$I_{sat,be}$	saturation current of base-emitter diode
I_{t1}	trigger current
I_{t2}	failure threshold current
I_{tr2}	trigger current base pushout
J	current density
J_{ideal}	ideal diode current density
$J_n, \vec{J}_n, J_e$	current density of electrons
$J_n, \vec{J}_n, J_h$	current density of holes
J_s	saturation current density (diodes)
k_B	Boltzmann's constant ($= 1.38 \times 10^{-23}$ J/K)
l	mean free path of carriers
L	design parameter for diodes, resistor length
l_g	gate length
$l_{g,eff}$	effective gate length
L_n (L_p)	diffusion length of electrons (holes)
L_s	series inductance in the ESD models
L_{th}	thermal diffusion length
M	Miller multiplication factor
m_0	effective electron mass in silicon
n	carrier density, electron density
n	ideality factor (emissivity) of diodes

n_{bc}	ideality factor (emissivity) of base-collector diode
n_{be}	ideality factor (emissivity) of base-emitter diode
N_{C}	collector background doping level
N_{D} (N_{A})	donor (acceptor) doping concentration
N_{e}	emitter doping concentration
n_{i}	intrinsic electron density
n_{p0}	equilibrium electron density on the p side (of a pn junction)
p	hole density
P_{n} (P_{p})	thermoelectric power of electrons (holes)
p_{n0}	equilibrium hole density on the n side (of a pn junction)
q	electronic charge (1.6×10^{-19} C)
Q_{b}	Gummel number
Q_{B}	charge stored in base of BJT (Gummel–Poon model)
Q_{B0}	base charge under zero bias of BJT (Gummel–Poon model)
Q_{dC} (Q_{dE})	charge stored in the diffusion capacitances of collector (emitter) in a BJT (Gummel–Poon model)
Q_{jC} (Q_{jE})	charge stored in the depletion region of collector (emitter) in a BJT (Gummel–Poon model)
R	recombination rate
R_{arc}	arc resistance during a CDM event
R_{b}	diffusion resistance of the base (in compact models)
R_{bd}	differential resistance in the breakdown branch
R_{c}	diffusion resistance of the collector (in compact models)
R_{CDM}	CDM resistor
R_{D}	serial resistance (in diode)
R_{diff}	differential resistance in the high-current regime
R_{e}	diffusion resistance of the emitter (in compact models)
R_{ESD}	ESD resistor (equivalent to R_{HBM}, R_{MM}, ..., depending on the model)
R_{HBM}	HBM resistor (1.5 kΩ for component-level tests)
R_{in}	resistor before 2nd stage in input cells
R_{L}	resistance of the load under ESD conditions
R_{MM}	MM resistor (0 Ω for component-level tests)
R_{SDM}	SDM resistor
R_{sub}	bulk (substrate) resistance
R_{sq}	sheet resistance
R_{th}	thermal resistances
R_{w}	well resistance (in compact models)
R_{wsb}, R_{wbs}	well resistance between base and source (in compact models)
SCG	source contact-to-gate spacing
SWS	source-to-well spacing
t	time

T	temperature
T_{crit}	critical (local) temperature in a device causing fail
$T_{\mathrm{e}}\ (T_{\mathrm{h}})$	carrier temperature of electrons (holes)
$t_{\mathrm{f}}\ (t_{\mathrm{r}})$	forward (reverse) base-transit time
t_{pulse}	pulse width
t_{rise}	rise time of ESD pulse
V	voltage
V_{bd}	breakdown voltage
V_{bc}	base-collector voltage
V_{be}	base-emitter voltage
V_{CDM}	CDM discharge voltage prior to discharge
V_{CE0}	open-base collector-emitter breakdown voltage
V_{device}	voltage across the DUT during an ESD stress
V_{ds}	drain-source voltage
V_{gs}	gate-source voltage
$v_{\mathrm{d}},\ \vec{v_{\mathrm{d}}}$	drift velocity of carriers
V_{ESD}	ESD discharge voltage prior to discharge
V_{gate}	gate potential
V_{h}	holding voltage
V_{HBM}	HBM discharge voltage prior to discharge
V_{MM}	MM discharge voltage prior to discharge
V_{on}	on-voltage of diodes/diode strings
v_{sat}	saturation velocity of carriers
V_{SDM}	SDM discharge voltage prior to discharge
V_{sig}	signal voltage
V_{t1}	trigger voltage
V_{t2}	voltage at failure current I_{t2}
V_{tr2}	trigger voltage base pushout
W	device width (perpendicular to current flow)
$w_{\mathrm{p+}}$	design parameter for diodes
α	damping coefficient (in discharge waveform)
$\alpha\ (\alpha_{\mathrm{n}},\ \alpha_{\mathrm{p}})$	ionization coefficient (of electrons, holes)
α_{f}	forward current gain of common-base BJT
α_{r}	reverse current gain of common-base BJT
β	current gain
β_{eff}	effective current gain
β_{f}	forward current gain of common-emitter BJT
β_{r}	reverse current gain of common-emitter BJT
ϵ	permittivity
κ	thermal conductivity
λ	wavelength
μ_0	low-field mobility

μ_n (μ_p)	electron (hole) mobility
ρ	charge
τ_max	maximum carrier lifetime
τ_n (τ_p)	lifetime of electrons (holes)
ω_0	oscillation frequency (in discharge waveform)

Abbreviations

ADC	analog to digital converter
AFM	atomic force microscopy
BiCMOS	bipolar / complementary metal-oxide semiconductor technology
BLI	backside laser interferometry
CBM	Charged Board Model
CDM	Charged Device Model
CFM	capacitive force microscopy
CMOS	complementary metal-oxide semiconductor
CMP	chemical mechanical polishing
CVD	chemical vapor deposition
DAC	digital to analog converter
DUT	device under test
EM	Ebers-Moll (model)
EMMI	emission microscopy
EOS	electric over-stress
ESD	electrostatic discharge
ESDA	Electrostatic Discharge Association
FD-SOI	fully depleted silicon on insulator
FET	field-effect transistor
FIB	focused ion beam
FWHM	full width at half maximum
gcNFET	gate-coupled n channel field-effect transistor
gds	graphic design system
ggNFET	grounded-gate n channel field-effect transistor
GOX	gate oxide
GP	Gummel-Poon (model)
HBM	Human Body Model
HDD	highly doped diffusion (of drain/source)
hf	high frequency
HV	high voltage
IC	integrated circuit
I^2C	inter-integrated-circuit bus
IMOX	inter-metal oxide
I/O	input/output
LDD	lightly doped drain
LEM	lumped element model
LOCOS	local oxidation of silicon
LVTSCR	low-voltage triggering silicon controlled rectifier
MM	Machine Model
MOSFET	metal oxide semiconductor field-effect transistor

NFET	n channel field-effect transistor
PCB	printed circuit board
PFET	p channel field-effect transistor
PLL	phase-locked loop
POR	process of record
rf	radio frequency
RIE	rapid ion etching
RSCE	reverse short channel effect
RTA	rapid thermal annealing
SCR	silicon controlled rectifier, thyristor
SDM	Socketed Device Model
SEM	scanning electron microscopy
SIC	Selectively Implanted Collector
SIMS	secondary ion mass spectroscopy
SOI	silicon on insulator
SRH	Shockley-Read-Hall (generation/recombination/mechanism)
STI	shallow trench isolation
TEM	transmission electron microscopy
TLP	Transmission Line Pulsing
VLSI	very large scale integration

Index

Colour section

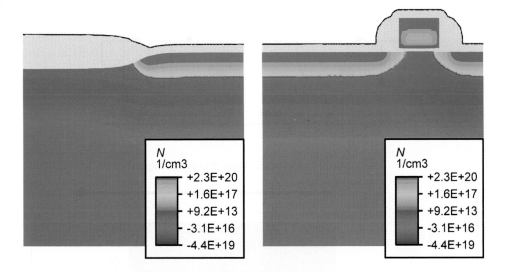

Figure 4.5: Left: LOCOS isolation showing the so called bird's beak. Right: Poly gate complex for an NFET device. At both positions, the resulting *pn* junction of drain to well shows comparable curvature (see page 81).

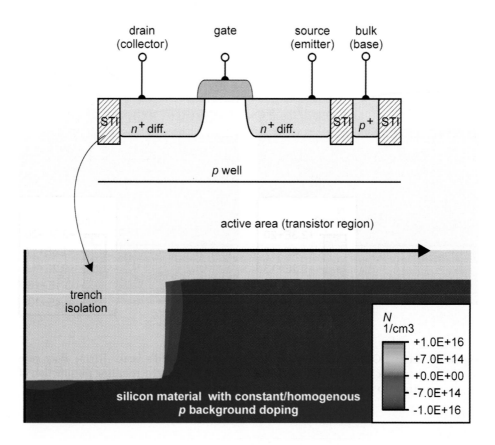

Figure 4.7: Lateral isolation of an active device by shallow trench isolation (see page 86).

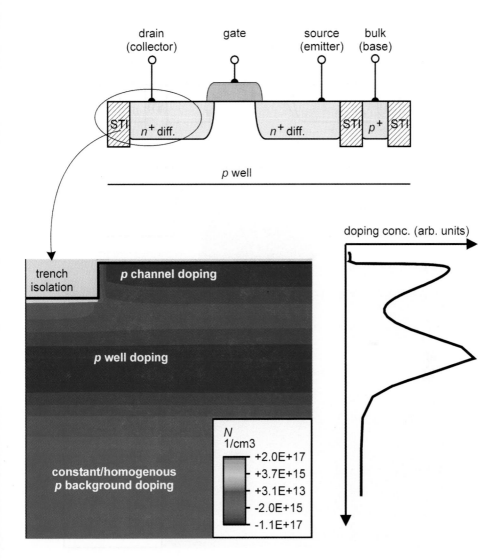

Figure 4.8: Implant of well and channel doping to control threshold and analogue behaviour (see page 87).

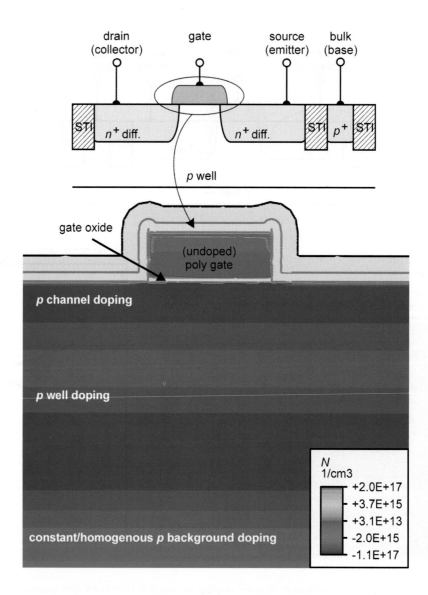

Figure 4.9: Deposition of poly silicon and structuring to form gate stack (see page 89).

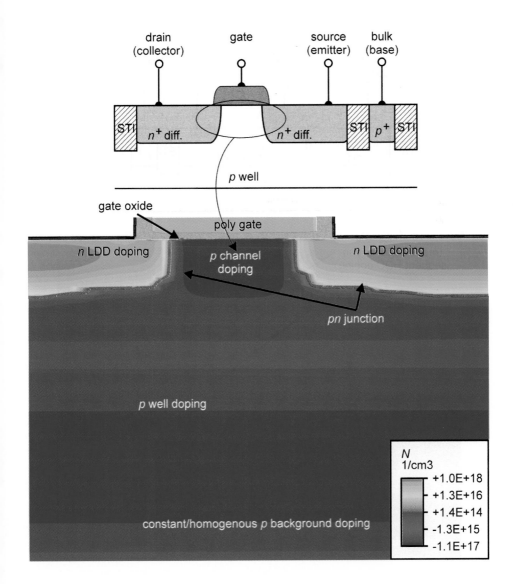

Figure 4.10: Implantation of source/drain extension (see page 90).

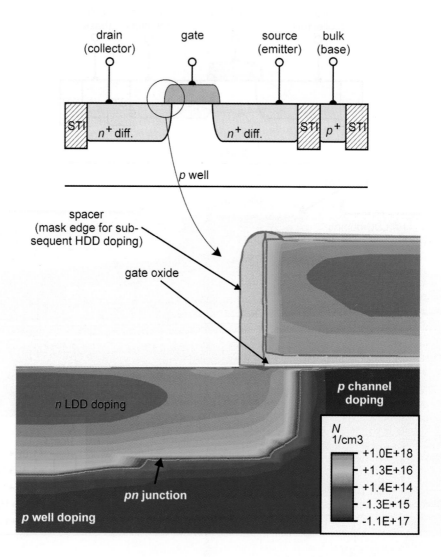

Figure 4.11: Deposition of poly silicon and structuring to form a gate stack (see page 91).

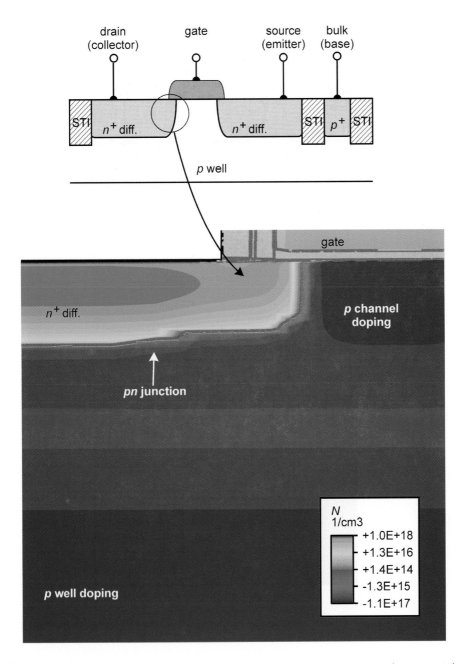

Figure 4.12: HDD implantation to reduce the parasitic diffusion resistance (see page 93).

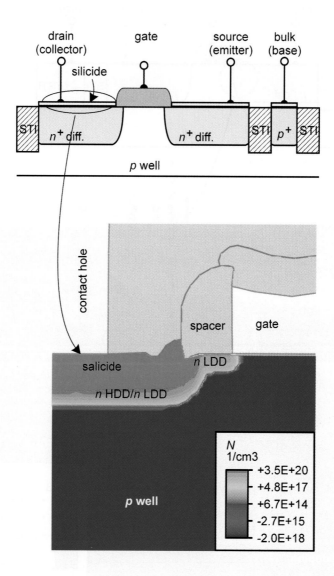

Figure 4.13: Silicide formation to reduce contact resistance (see page 94).

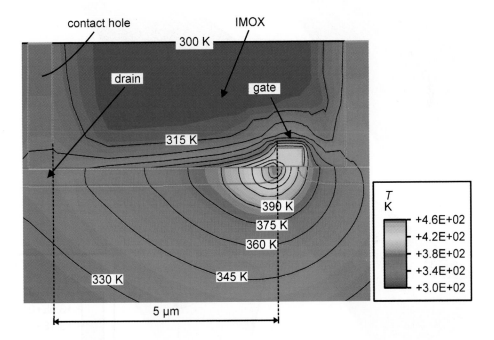

Figure 5.11: Temperature distribution in a ggNMOS 150 ns after the beginning of a TL
pulse of 8 mA/µm. The isothermal lines are plotted with a spacing of 15 K
(see page 114).

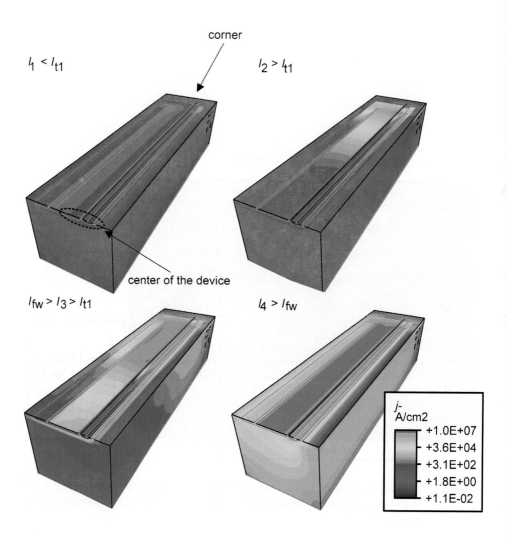

Figure 5.26: Current density distribution inside the ggNFET at different current levels. After snapback ($I_2 > I_{t1}$), the current is confined to a region at the device corners and spreads out for higher currents, until the entire device is triggered at I_{fw} (see page 131).

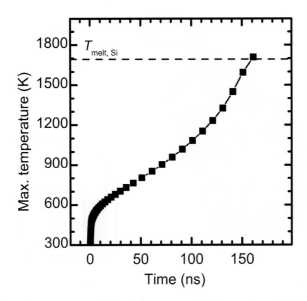

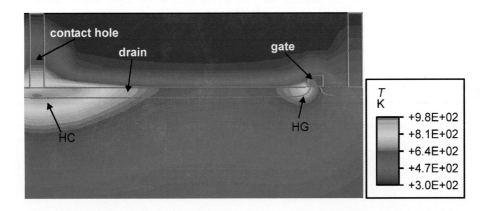

Figure 5.33: Transient evolution of the maximum temperature inside the NFET for a current pulse of $I_{TLP} = 26$ mA/μm (top). The temperature exceeds the melting point of silicon after 160 ns. The simulated temperature distribution for the NFET reveals a hot spot HC under the contact holes due to a vertical electric breakdown of the drain-to-substrate pn junction at this stress level (bottom). The hot spot HG is due to the laterally triggered parasitic bipolar transistor of the NFET and for the discussed stress condition has a lower temperature during the whole pulse duration (see page 140).

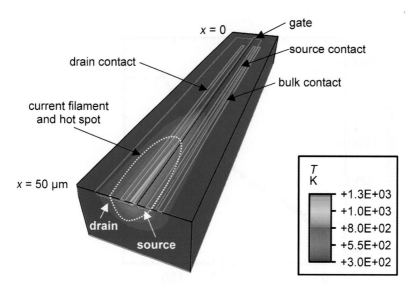

Figure 5.40: Temperature distribution for the ggNFET with minimum DCG showing a
destructive hot spot in the middle of the device. The picture shows a I_{TLP}
of 0.1 A. Because of insufficient ballasting resistance, the intrinsic current
density in the filament for that stress level is too high, which causes high
power dissipation in a confined volume, strong temperature increase and,
finally, melting of the silicon (see page 147).

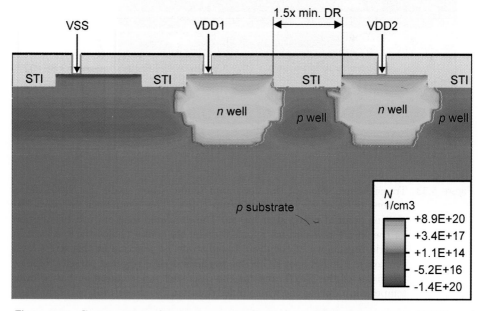

Figure 5.41: Cross section showing two n wells tied to different potentials VDD1 and
VDD2, representing a parasitic device with a specific breakdown voltage
between both power supply voltages (see page 148).

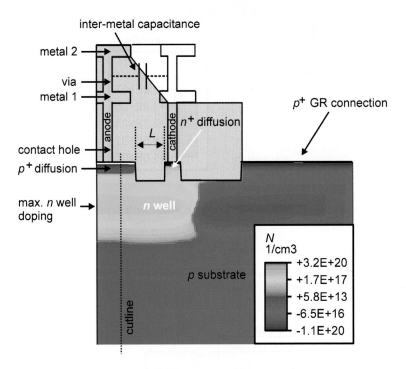

Figure 5.46: Topology and doping profile distribution of an STI-bound p^+ in n well diode used for 2D device simulation (see page 155).

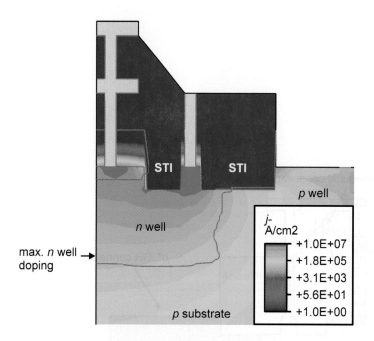

Figure 5.48: Current distribution in an STI bounded diode with $L = 1$ μm under forward
biased high current injection conditions (see page 157).

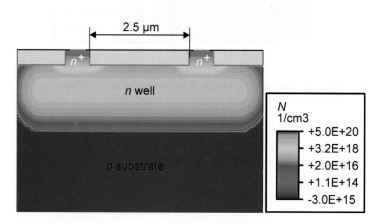

Figure 5.51: Cross section of an n well resistor of 2.5 μm length in a p substrate (see
page 159).

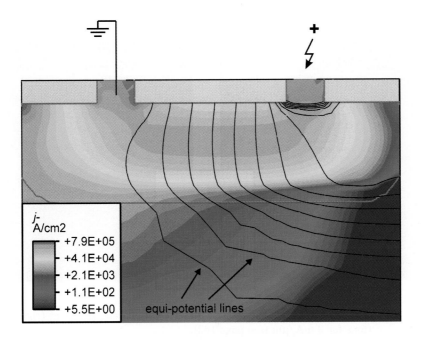

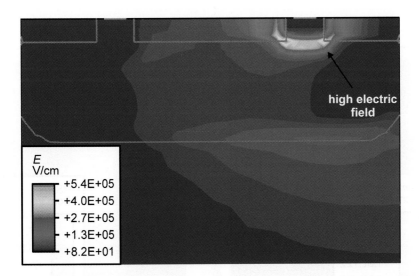

Figure 5.53: Top: Current flow and equipotential lines for the n well resistor under bias conditions. Bottom: Distribution of the electric field inside the structure prior to snapback. As the electric field increases at the n well n^+ contact of the hot pin with increasing voltage, there is an increase in the electron/hole generation, due to avalanche, which finally gives rise to the snapback of the device (see page 161).

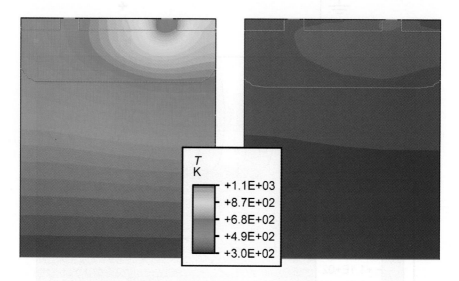

Figure 5.54: Temperature inside the resistor under DC (left) and transient (right) conditions for 2 mA/µm (see page 162).

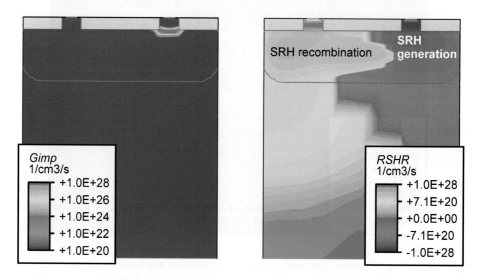

Figure 5.55: Impact ionization (left) and (SRH)-thermal generation (right) at snapback under DC conditions for 2 mA/µm. Here the thermally induced carriers cause snapback (see page 163).

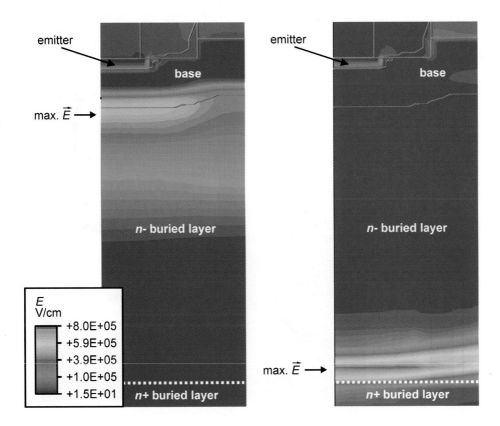

Figure 5.58: Simulation of the electric field distribution E before (left) and after (right) base pushout. The distribution of the maximum electric field moves from the pn junction of the base–collector junction to higher doped collector sections. A selectively implanted collector (SIC) has been used (see also Figure 5.56 and page 166).

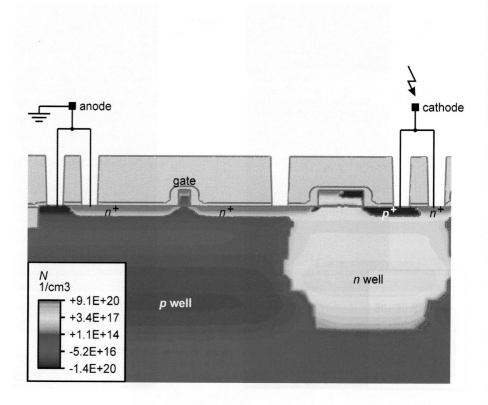

Figure 5.60: Cross-section showing the doping profile of a LVTSCR. (see page 168).

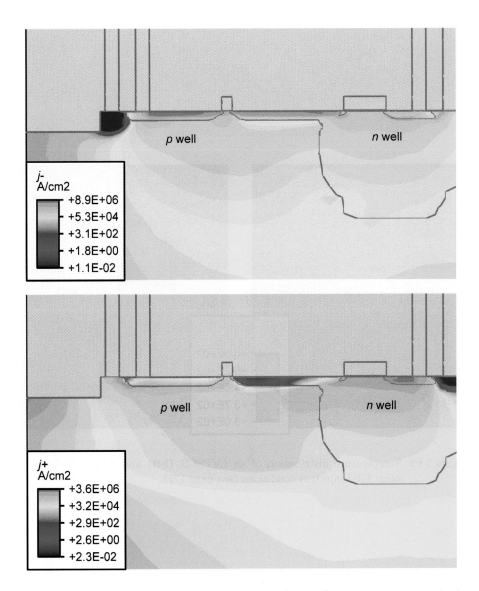

Figure 5.62: Distribution of electron (top) and hole (bottom) current density inside the LVTSCR under ESD stress conditions indicating *npn* and *pnp* bipolar operation of the device (see page 170).

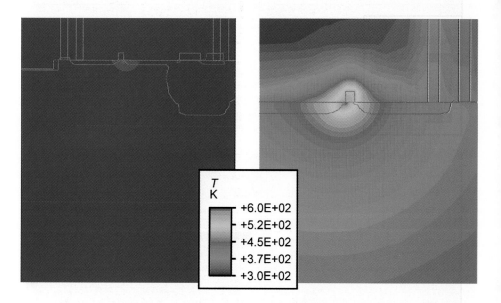

Figure 5.63: Temperature distribution of an LVTSCR (left) and NFET (right) under same ESD injection conditions (see page 171).

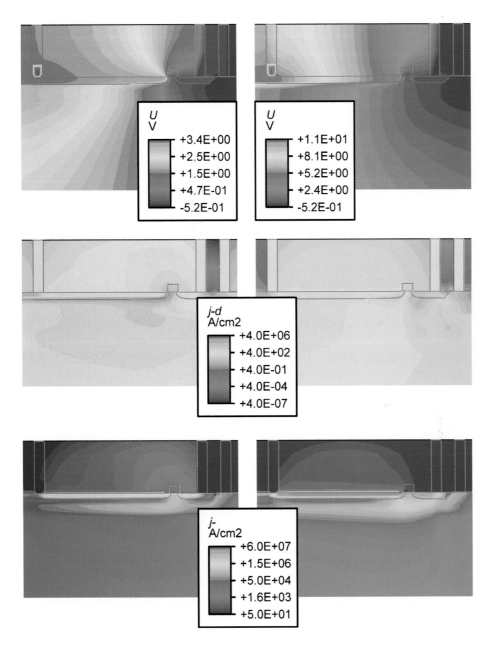

Figure 5.76: Electrostatic potential (top) displacement current density (middle) and total current density (bottom) distribution of the ggNFET with gate length
0.35 μm before, $t = 10$ ps, (left) and after, $t = 50$ ps, (right) triggering of
the parasitic bipolar transistor for a CDM stress of 500 V (see page 184).